Fundamentals of Massive MIMO

Written by pioneers of the concept, this is the first complete guide to the physical and engineering principles of Massive MIMO. Assuming only a basic background in communications and statistical signal processing, it will guide readers through key topics in multi-cell systems such as propagation modeling, multiplexing and de-multiplexing, channel estimation, power control, and performance evaluation. The authors' unique capacity-bounding approach will enable readers to carry out effective system performance analyses and develop advanced Massive MIMO techniques and algorithms. Numerous case studies, as well as problem sets and solutions accompanying the book online, will help readers put knowledge into practice and acquire the skill set needed to design and analyze complex wireless communication systems. Whether you are a graduate student, researcher, or industry professional working in the field of wireless communications, this will be an indispensable guide for years to come.

Thomas L. Marzetta is the originator of Massive MIMO. He is a Bell Labs Fellow at Nokia and a Fellow of the IEEE, and previously worked at Schlumberger-Doll Research and Nichols Research Corporation. He has received numerous recognitions and awards, including the IEEE W. R. G. Baker Award (2015), the IEEE Communications Society Stephen O. Rice Prize (2015), and the IEEE Guglielmo Marconi Prize Paper Award (2013). He is the recipient of an Honorary Doctorate from Linköping University.

Erik G. Larsson is a Professor at Linköping University in Sweden and a Fellow of the IEEE. He is the co-author of 120 journal papers and the textbook *Space-Time Block Coding for Wireless Communications* (Cambridge, 2003). He currently serves as Chair of the IEEE SPS Technical Committee for Signal Processing for Communications and Networking (2015–2016). In both 2012 and 2014, he received the IEEE Signal Processing Magazine Best Column Award, and in 2015 he was awarded the IEEE Communications Society Stephen O. Rice Prize.

Hong Yang is a member of technical staff at Nokia Bell Labs' Mathematics of Networks and Communications Research Department, where he conducts research into wireless communications networks. He has over fifteen years of industrial research and development experience, and has previously worked in both the Department of Radio Frequency Technology Systems Engineering and the Wireless Design Center at Alcatel-Lucent.

Hien Quoc Ngo is a researcher at Linköping University, Sweden. In 2015, he received the IEEE Communications Society Stephen O. Rice Prize and the IEEE Sweden VT-COM-IT Joint Chapter Best Student Journal Paper Award. He was also an IEEE Communications Letters exemplary reviewer in 2014 and an IEEE Transactions on Communications exemplary reviewer in 2015.

"Massive MIMO has over the past few years become one of the hottest research topics in wireless, and will be a key component of 5G. This book is written by pioneers of the area in a systematic and lucid way, and works out the fundamentals without getting lost in the details. I highly recommend it to anybody working in this field."

Andreas Molisch, *University of Southern California*

"Bright and profound, this book provides the fundamentals to understand the unique capabilities of Massive MIMO and illustrates the benefits for specific use cases. The authors are scientific pioneers and masters in explaining and sharing their proficiency in this book: it is an intellectual treat for everyone fascinated by Massive MIMO technology!"

Liesbet Van der Perre, *KU Leuven*

"A very timely text by some of the 'founding fathers' of massive MIMO. This is a great book for the beginner, with its simple but enlightening examples, as well as a great reference text for the more experienced engineer. The book is concise and to the point, and the summary and key points at the end of each chapter make it easy to focus your reading. Highly recommended for those that want to get an in-depth understanding of massive MIMO without spending months doing so."

Lee Swindlehurst, *University of California, Irvine*

Fundamentals of Massive MIMO

Thomas L. Marzetta
Bell Labs, Nokia

Erik G. Larsson
Linköping University, Sweden

Hong Yang
Bell Labs, Nokia

Hien Quoc Ngo
Linköping University, Sweden

CAMBRIDGE
UNIVERSITY PRESS

CAMBRIDGE
UNIVERSITY PRESS

University Printing House, Cambridge CB2 8BS, United Kingdom

One Liberty Plaza, 20th Floor, New York, NY 10006, USA

477 Williamstown Road, Port Melbourne, VIC 3207, Australia

314-321, 3rd Floor, Plot 3, Splendor Forum, Jasola District Centre, New Delhi - 110025, India

79 Anson Road, #06-04/06, Singapore 079906

Cambridge University Press is part of the University of Cambridge.

It furthers the University's mission by disseminating knowledge in the pursuit of education, learning and research at the highest international levels of excellence.

www.cambridge.org
Information on this title: www.cambridge.org/9781107175570
10.1017/9781316799895

First published 2016
Reprinted 2017

A catalogue record for this publication is available from the British Library

Library of Congress Cataloging in Publication data
Names: Marzetta, Thomas L., author.
Title: Fundamentals of massive MIMO / Thomas L. Marzetta, Bell Labs, Nokia [and 3 others].
Description: New York : Cambridge University Press, 2016.
Identifiers: LCCN 2016031164 | ISBN 9781107175570 (Hardback)
Subjects: LCSH: MIMO systems. | Wireless communication systems. | BISAC: TECHNOLOGY & ENGINEERING / Mobile & Wireless Communications.
Classification: LCC TK5103.4836 .M37 2016 | DDC 621.3841–dc23
LC record available at https://lccn.loc.gov/2016031164

ISBN 978-1-107-17557-0 Hardback

Additional resources for this title are available at www.cambridge.org/Marzetta

Contents

Figures

Tables

PREFACE

Simplification of modes of proof is not merely an indication of advance in our knowledge of a subject, but is also the surest guarantee of readiness for further progress.

– W. Thomson [1st Baron Kelvin] and P. G. Tait,
Elements of Natural Philosophy, 1873

There are three timeless truths in the field of wireless communications:

- Demand for wireless throughput, both mobile and fixed, will always increase.
- The quantity of available electromagnetic spectrum will never increase, and the most desirable frequency bands that can propagate into buildings and around obstacles and that are unaffected by weather constitute only a small fraction of the entire spectrum.
- Communication theorists and engineers will always be pressured to invent or to discover breakthrough technologies that provide higher spectral efficiency.

Given the history of more than a century of wireless innovation, we must look to the physical layer for breakthrough technologies. The central assertion of this book is that Massive MIMO constitutes a breakthrough technology. It is a scalable technology whereby large numbers of terminals simultaneously communicate through the entire allocated frequency spectrum. What enables this aggressive multiplexing is, first, an excess number of service antennas compared with terminals, and, second, performing the multiplexing and de-multiplexing based on *measured* propagation characteristics rather than on *assumed* propagation characteristics. A disproportionate number of service antennas compared with active terminals makes it likely that the propagation channels are conducive to successful multiplexing, and basing the multiplexing on direct channel measurements makes the antenna array tolerances independent of the number of antennas. The activity of growing the number of antennas relative to the number of active terminals renders the simplest types of multiplexing and de-multiplexing signal processing exceedingly effective, and it permits the same quality of service with reduced radiated power. Low radiated power is conducive to frequency reuse. The combined action of many antennas eliminates frequency-dependent fading and simplifies power control. By virtue of the time-division duplexing operation, the propagation channel characteristics are measured on the uplink and used both for uplink data detection and downlink beamforming. This facilitates operation in high-mobility scenarios. Also, by performing appropriate power control, Massive MIMO yields uniformly good service to all terminals, as measured in terms of 95%-likely throughputs on both the uplink and downlink.

Ostensibly, analyzing the performance of a Massive MIMO system is a daunting task because of the sheer number of frequency dependent propagation channels at work, and the fact that all terminals transmit and receive information over all frequencies. The central message of this book is that substantially closed-form performance expressions are obtainable for even the most complicated multi-cell Massive MIMO deployments. We achieve analytical tractability in three ways: first, we model small-scale fading – a-priori unknown to everyone – as independent, Rayleigh distributed; second, we assume that large-scale fading is known to everyone who needs to know it; and, third, we restrict attention to the simplest linear multiplexing and de-multiplexing – zero-forcing and maximum-ratio processing, both on the uplink and on the downlink. Collectively, these assumptions admit Bayesian analysis and ergodic capacity lower bounds, whose derivation requires only elementary mathematical techniques. For multi-cell Massive MIMO deployment, we obtain comprehensive but remarkably simple non-asymptotic expressions for the capacity lower bounds. Massive MIMO, in effect, creates a dedicated virtual circuit between the home base station and each of its terminals, comprising a frequency independent channel whose quality depends only on large-scale fading and power control. Our capacity lower bounds account for receiver noise, channel estimation errors, the overhead associated with pilots, power control, the imperfections of the particular multiplexing or de-multiplexing signal processing that is employed, non-coherent inter-cell interference, and coherent inter-cell interference that arises from pilot reuse. These bounds yield considerable intuitive insight into the workings of Massive MIMO systems, the interplay of system parameters, and system scalability. Numerical case studies illustrate the tremendous potential of Massive MIMO as well as the value of the capacity bounds as system design tools.

This book should appeal to three classes of readers. Wireless engineers will find a clear exposition of the principles of Massive MIMO that uses only elementary communication theory and statistical signal processing. While exhaustive end-to-end system simulations may be employed for detailed system performance analyses, and possibly for engineering designs, our capacity-bound approach is likely to see extensive use because of its speed and simplicity, the insight that it gives into the interaction of system parameters, and as an independent check on simulations. Researchers who are devising advanced Massive MIMO techniques and algorithms can use baseline performance measures, embodied in our capacity bounds, to quantify performance improvements versus implementation complexity. For the student, this book should be an ideal vehicle for learning how to translate basic information theory, and communication and signal processing principles, into the analysis of complicated communication systems.

How to Read This Book

With the aid of the appendices, this book is self-contained and requires only linear algebra and undergraduate-level probability theory as prerequisites.

Practicing systems engineers looking for a quick insight into the potential of Massive MIMO technology may start by reading Chapter 6 on case studies, and then for insights into the performance evaluation methodology, read Chapter 3 for the single-cell analysis, and Chapter 4 for the multi-cell analysis. The other chapters and appendices can be used as references when needed.

Professors looking to cover the topic in depth (a one-semester graduate-level course), or students and researchers looking for a solid background in Massive MIMO performance analysis, may first read Chapters 1–2, next study the background material in Appendices A–C, and then read Chapters 3–8 in sequence, referring to the rest of the appendices whenever needed.

A set of problems to each chapter, and an accompanying solution manual, are available to course instructors and may be obtained by contacting the authors.

Chapter 1

INTRODUCTION

The performance limitation of any wireless network will always be at the physical layer, because, fundamentally, the amount of information that can be transferred between two locations is limited by the availability of spectrum, the laws of electromagnetic propagation, and the principles of information theory.

There are three basic ways in which the efficiency of a wireless network may be improved: (i) deploying access points more densely; (ii) using more spectrum; and (iii) increasing the *spectral efficiency*, that is, the number of bits that can be conveyed per second in each unit of bandwidth. While future wireless systems and standards are likely to use an ever-increasing access point density and use new spectral bands, the need for maximizing the spectral efficiency in a given band is never going to vanish.

The use of multiple antennas, also known as *multiple-input, multiple-output* (MIMO) technology, is the only viable approach for substantial improvement of spectral efficiency. While mostly developed during the last two decades, it is noteworthy that a basic idea behind MIMO is almost a century old: in [1], directional beamforming using an antenna array was suggested to permit more aggressive frequency reuse of scarce spectrum – in this case, very low frequency – for transoceanic communication.

MIMO technology is logically classified into one of three categories, whose development occurred during roughly disjoint epochs: Point-to-Point MIMO, Multiuser MIMO, and Massive MIMO. This book is about Massive MIMO, which arguably will be the ultimate embodiment of MIMO technology. The following sections explain these incarnations of MIMO and their important differences. This treatment is intended to be a quick overview, and subsequent chapters will expand upon the concepts introduced here.

1.1 Point-to-Point MIMO

Point-to-Point MIMO emerged in the late 1990s [2–11] and represents the simplest form of MIMO: a base station equipped with an antenna array serves a terminal equipped with an antenna array; see Figure 1.1. Different terminals are orthogonally multiplexed, for example via a combination of time- and frequency-division multiplexing. In what follows, we summarize some basic facts about Point-to-Point MIMO. More details, along with derivations of all formulas given here, are provided in Section C.3.

In each channel use, a vector is transmitted and a vector is received. In the presence of additive white Gaussian noise at the receiver, Shannon theory yields the following formulas for the link spectral efficiency (in b/s/Hz):

$$C^{\mathrm{ul}} = \log_2 \left| \boldsymbol{I}_M + \frac{\rho_{\mathrm{ul}}}{K} \boldsymbol{G}\boldsymbol{G}^{\mathrm{H}} \right|, \tag{1.1}$$

$$\begin{aligned} C^{\mathrm{dl}} &= \log_2 \left| \boldsymbol{I}_K + \frac{\rho_{\mathrm{dl}}}{M} \boldsymbol{G}^{\mathrm{H}}\boldsymbol{G} \right| \\ &\overset{(a)}{=} \log_2 \left| \boldsymbol{I}_M + \frac{\rho_{\mathrm{dl}}}{M} \boldsymbol{G}\boldsymbol{G}^{\mathrm{H}} \right|. \end{aligned} \tag{1.2}$$

In (1.1) and (1.2), $\boldsymbol{G}$ is an $M \times K$ matrix that represents the frequency response of the channel between the base station array and the terminal array; ρ_{ul} and ρ_{dl} are the uplink and downlink signal-to-noise ratios (SNRs), which are proportional to the corresponding total radiated powers; M is the number of base station antennas; and K is the number of terminal antennas. Also, in (a) we used Sylvester's determinant theorem. The normalization by K and M reflects the fact that for constant values of ρ_{ul} and ρ_{dl} total radiated power is independent of the number of antennas. The spectral efficiency values in (1.1) and (1.2) require the receiver to know $\boldsymbol{G}$ but do not require the transmitter to know $\boldsymbol{G}$. Performance can be improved somewhat if the transmitter also acquires channel state information (CSI). However, this requires special effort and is seldom seen in practice – see Section C.3 for the associated capacity formula.

In isotropic (rich) scattering propagation environments, well modeled by independent Rayleigh fading, for sufficiently high SNRs, C^{ul} and C^{dl} scale linearly with $\min(M, K)$ and logarithmically with the SNR. Hence, in theory, the link spectral efficiency can be increased by simultaneously using large arrays at the transmitter and the receiver, that is, making M and K large. In practice, however, three factors seriously limit the usefulness of Point-to-Point MIMO, even with large arrays at both ends of the link. First, the terminal equipment is complicated, requiring independent RF chains per antenna as well as the use of advanced digital processing to separate the data streams. Second, more fundamentally, the propagation environment must support $\min(M, K)$ independent streams. This is often not the case in practice when compact arrays are used. Line-of-sight (LoS) conditions are

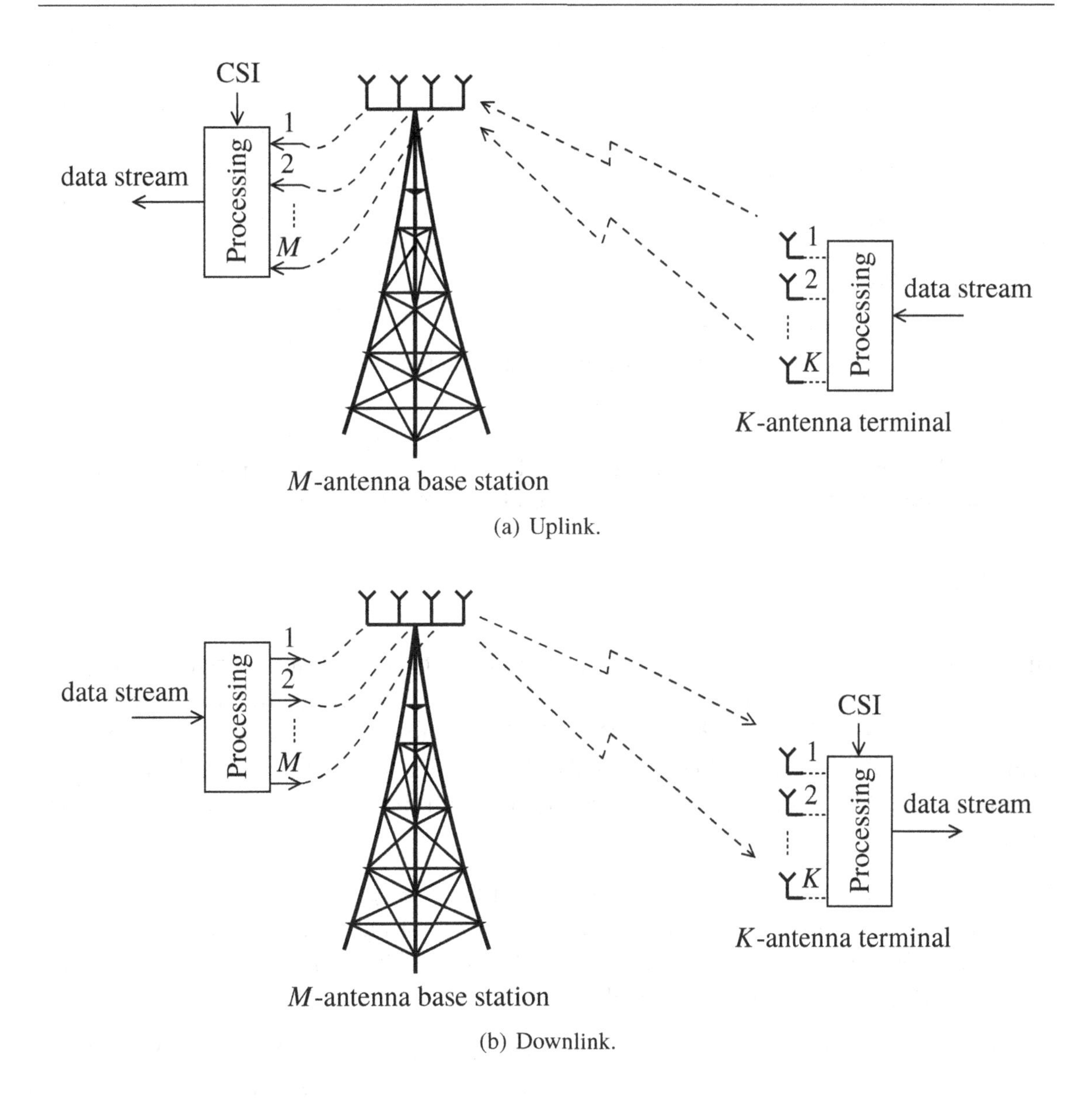

(a) Uplink.

(b) Downlink.

Figure 1.1. Point-to-Point MIMO.

particularly stressing. Third, near the cell edge, where normally a majority of the terminals are located and where SNR is typically low because of high path loss, the spectral efficiency scales slowly with $\min(M, K)$. Figure 1.2 illustrates this problem on the downlink for a terminal with $K = 4$ antennas and an SNR of -3 dB.

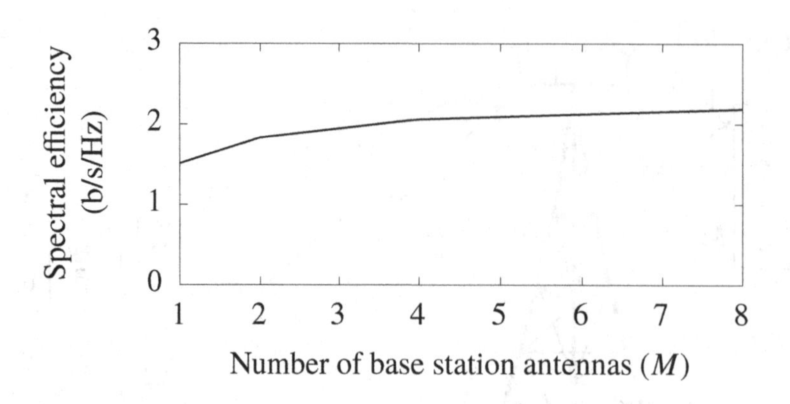

Figure 1.2. Downlink spectral efficiency with Point-to-Point MIMO for a terminal at the cell edge with $K = 4$ antennas, no CSI at the base station, and an SNR of −3 dB.

1.2 Multiuser MIMO

The idea of Multiuser MIMO is for a single base station to serve a multiplicity of terminals using the same time-frequency resources; see Figure 1.3. Effectively, the Multiuser MIMO scenario is obtained from the Point-to-Point MIMO setup by breaking up the K-antenna terminal into multiple autonomous terminals. This section summarizes some basic results of Multiuser MIMO. More details, and derivations of all formulas stated here, are given in Section C.4.

The basic concept of serving several terminals simultaneously using an antenna array at the base station is quite old [12–19]. However, a rigorous information-theoretic understanding of Multiuser MIMO emerged much later [20–23]. The transition in thinking from Point-to-Point MIMO to Multiuser MIMO is explained in some detail in [24].

Our discussion in this section is confined to that particular form of Multiuser MIMO for which there is a comprehensive Shannon theory which provides the ultimate performance of the system and specifies how this performance may be approached arbitrarily closely. It will be convenient to call this *conventional* Multiuser MIMO, even though it is doubtful if such a system has ever been reduced to practice.

Throughout this book, we assume that terminals in Multiuser MIMO have a single antenna. Hence, in the setup in Figure 1.3 the base station serves K terminals. Let $\boldsymbol{G}$ be an $M \times K$ matrix corresponding to the frequency response between the base station array and the K

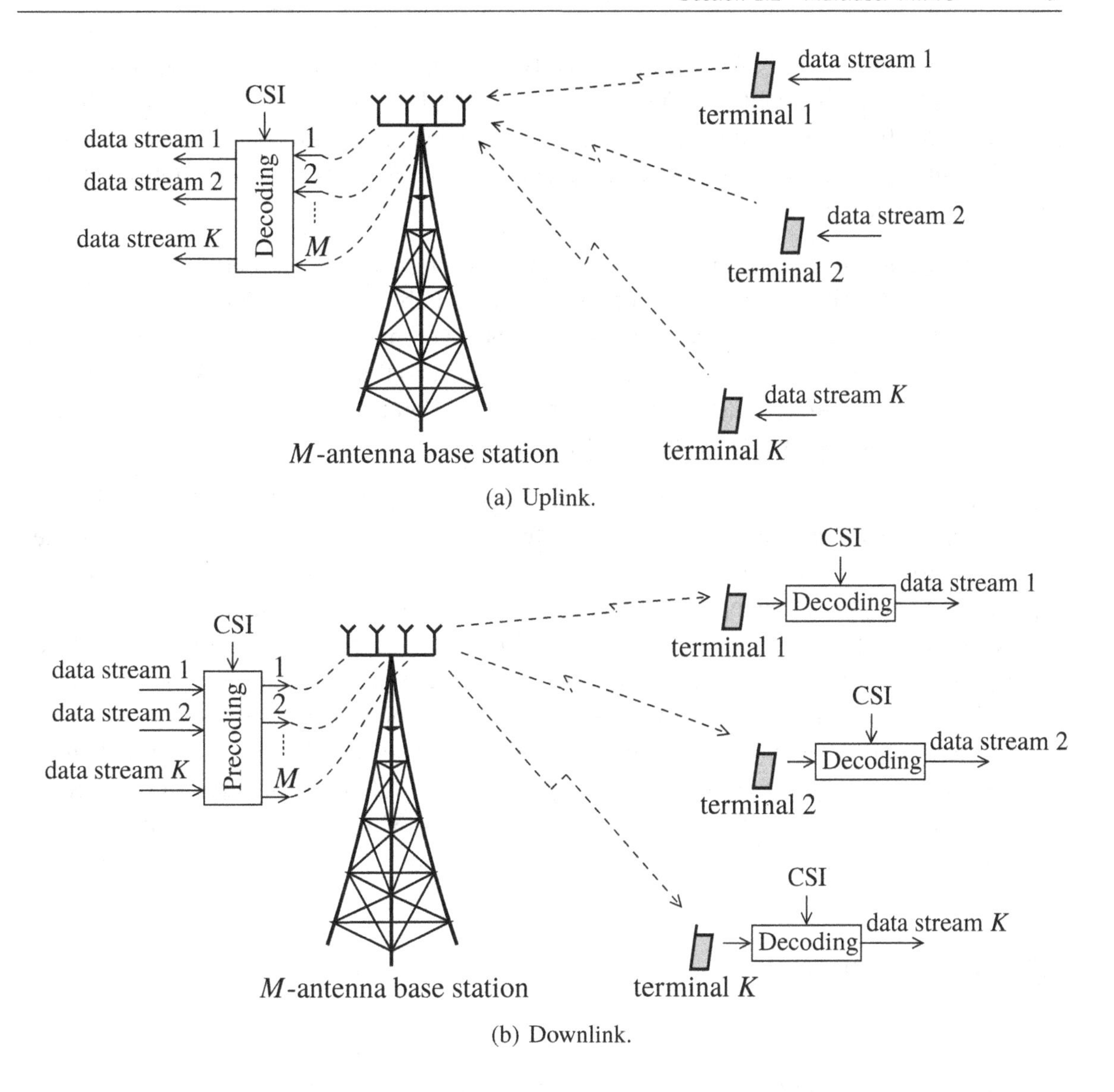

Figure 1.3. Multiuser MIMO.

terminals. The uplink and downlink sum spectral efficiencies are given by

$$C^{\mathrm{ul}} = \log_2 \left| \boldsymbol{I}_M + \rho_{\mathrm{ul}} \boldsymbol{G}\boldsymbol{G}^{\mathrm{H}} \right|, \tag{1.3}$$

$$C^{\mathrm{dl}} = \max_{\substack{\nu_k \geq 0 \\ \sum_{k=1}^{K} \nu_k \leq 1}} \log_2 \left| \boldsymbol{I}_M + \rho_{\mathrm{dl}} \boldsymbol{G}\boldsymbol{D}_{\boldsymbol{\nu}}\boldsymbol{G}^{\mathrm{H}} \right|, \tag{1.4}$$

where $\boldsymbol{\nu} = [\nu_1, \ldots, \nu_K]^{\mathrm{T}}$, ρ_{ul} is the uplink SNR per terminal, and ρ_{dl} is the downlink SNR. (For given ρ_{ul}, the total uplink power is K times greater than for the Point-to-Point MIMO model.) The computation of downlink capacity according to (1.4) requires the solution of a convex optimization problem. The possession of CSI is crucial to both (1.3) and (1.4). On uplink, the base station alone must know the channels, and each terminal has to be told its permissible transmission rate separately. On the downlink, both the base station and the terminals must have CSI.

Note that the terminal antennas in the point-to-point case can cooperate, whereas the terminals in the multiuser case cannot. Quite remarkably, however, the inability of the terminals to cooperate in the multiuser system does not compromise the uplink sum spectral efficiency as seen by comparing (1.1) and (1.3). Note also that the downlink capacity (1.4) may exceed the downlink capacity in (1.2) for Point-to-Point MIMO, because (1.4) assumes that the base station knows $\boldsymbol{G}$, where as (1.2) does not.

Multiuser MIMO has two fundamental advantages over Point-to-Point MIMO. First, it is much less sensitive to assumptions about the propagation environment. For example, LoS conditions are stressing for Point-to-Point MIMO, but not for Multiuser MIMO, as explained in Chapter 7. Second, Multiuser MIMO requires only single-antenna terminals. Notwithstanding these virtues, two factors seriously limit the practicality of Multiuser MIMO in its originally conceived form. First, to achieve the spectral efficiencies in (1.3) and (1.4) requires complicated signal processing by both the base station and the terminals. Second, and more seriously, on the downlink both the base station and the terminals must know $\boldsymbol{G}$, which requires substantial resources to be set aside for transmission of pilots in both directions. For these reasons, the original form of Multiuser MIMO is not scalable either with respect to M or to K.

1.3 Massive MIMO

Originally conceived in [25,26], Massive MIMO is a useful and scalable version of Multiuser MIMO. This section introduces the basic Massive MIMO concepts.

Consideration of net spectral efficiency alone according to the rigorous Shannon theory that underlies (1.3) and (1.4) suggests the optimality of a rough parity between M and K in conventional Multiuser MIMO: further growth of M only yields logarithmically increasing throughputs while incurring linearly increasing amounts of time spent on training. Massive MIMO represents a clean break from conventional Multiuser MIMO. Measures are taken such that one operates farther from the Shannon limit, but paradoxically achieves much better performance than any conventional Multiuser MIMO system.

There are three fundamental distinctions between Massive MIMO and conventional Multiuser MIMO. First, only the base station learns $\boldsymbol{G}$. Second, M is typically much larger than K, although this does not have to be the case. Third, simple linear signal processing is used both on the uplink and on the downlink. These features render Massive MIMO *scalable* with respect to the number of base station antennas, M.

Figure 1.4 illustrates the basic Massive MIMO setup. Each base station is equipped with a large number of antennas, M, and serves a cell with a large number of terminals, K. The terminals typically (and throughout this book) have a single antenna each. Different base stations serve different cells, and with the possible exception of power control and pilot assignment, Massive MIMO uses no cooperation among base stations.

Either in uplink or in downlink transmissions, all terminals occupy the full time-frequency resources concurrently. On uplink, the base station has to recover the individual signals transmitted by the terminals. On the downlink, the base station has to ensure that each terminal receives only the signal intended for it. The base station's multiplexing and de-multiplexing signal processing is made possible by utilizing a large number of antennas and by its possession of CSI.

Under LoS propagation conditions, the base station creates, for each terminal, a beam within a narrow angular window centered around the direction to the terminal; see Figure 1.5(a). The more antennas, the narrower are the beams. By contrast, in the presence of local scattering, the signal seen at any given point in space is the superposition of many independently scattered and reflected components that may add up constructively or destructively. When the transmitted waveforms are properly chosen, these components superimpose constructively precisely at the locations of the terminals; see Figure 1.5(b). The more antennas, the more sharply the power focuses onto the terminals. When focusing the power, the use of sufficiently accurate CSI at the base station is essential. In time-division duplex operation (TDD), the base station acquires CSI by measuring pilots transmitted by the terminals, and exploiting reciprocity between the uplink and downlink channel. This requires reciprocity calibration of the transceiver hardware, as discussed in Section 8.7. However, phase-calibrated arrays are not required, since by virtue of the reciprocity a phase offset between any two antennas will affect the uplink and the downlink in the same way.

Increasing the number of antennas, M, always improves performance, in terms of both reduced radiated power and in terms of the number of terminals that can be simultaneously served. In Chapters 3 and 4, we give rigorous lower bounds on Massive MIMO spectral efficiency, and these bounds account for all overhead and imperfections associated with estimating the channels from uplink pilots.

The use of large numbers of antennas at the base station is instrumental not only to obtain high sum spectral efficiencies in a cell, but, more importantly, to provide uniformly good service to many terminals simultaneously. An additional consequence of using large numbers of antennas is that the required signal processing and resource allocation

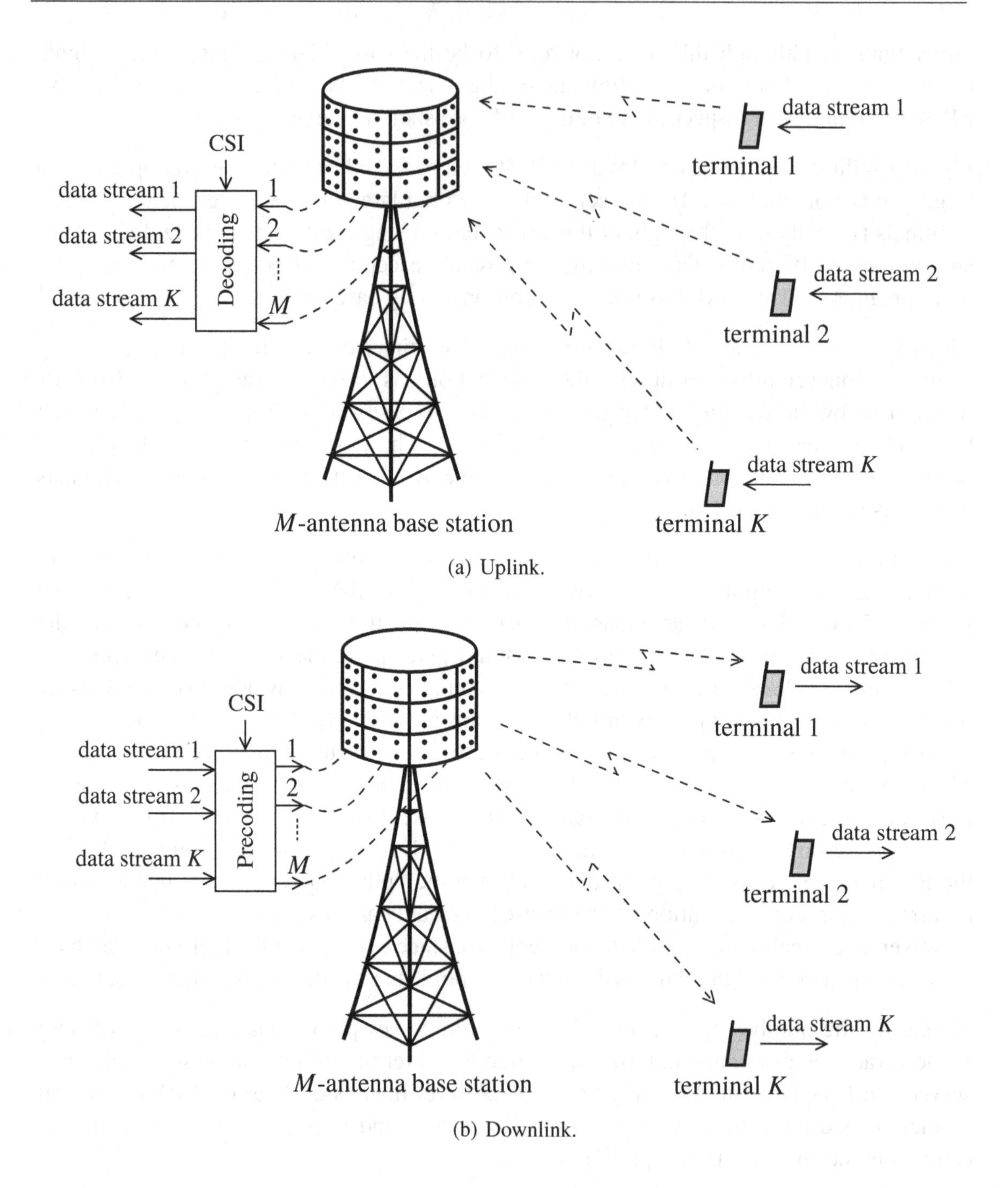

(a) Uplink.

(b) Downlink.

Figure 1.4. Massive MIMO.

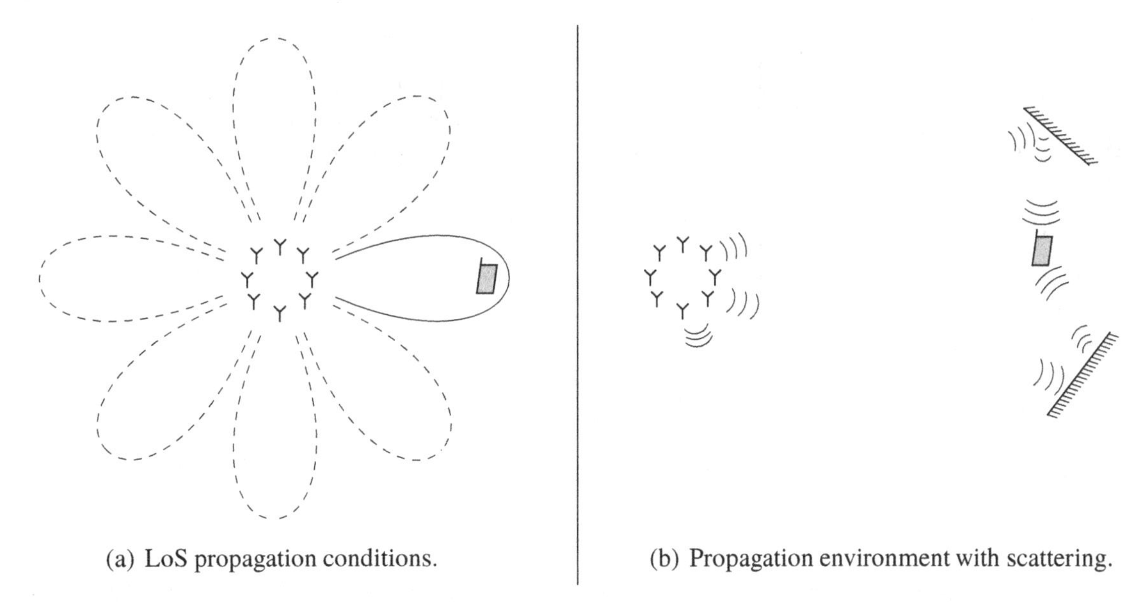

(a) LoS propagation conditions.

(b) Propagation environment with scattering.

Figure 1.5. The effect of precoding in different propagation environments.

simplifies, owing to a phenomenon known as *channel hardening*. The significance of channel hardening is that effects of small-scale fading and frequency dependence disappear when M is large. Specifically, consider a terminal with M-dimensional channel response $\boldsymbol{g}$; if beamforming with a beamforming vector $\boldsymbol{a}$ is applied, then the terminal sees a scalar channel with gain $\boldsymbol{a}^{\mathrm{T}}\boldsymbol{g}$. When M is large, by virtue of the law of large numbers, $\boldsymbol{a}^{\mathrm{T}}\boldsymbol{g}$ is close to its expected value, $\mathsf{E}\left\{\boldsymbol{a}^{\mathrm{T}}\boldsymbol{g}\right\}$ (a deterministic number). This means that the resulting *effective channel* between each terminal and the base station is a scalar channel with known, frequency-independent gain and additive noise. We show in Chapters 3 and 4 that the capacity of this channel can be rigorously, and without approximations, characterized in terms of an *effective signal-to-interference-plus-noise ratio (SINR)*. Importantly, this characterization does not rely on channel hardening and it is valid for any M and K; however, by virtue of channel hardening, most relevant capacity bounds are tight only when M is reasonably large. This characterization in turn facilitates the use of simple schemes for resource allocation and power control, as further explained in Chapter 5. Furthermore, channel hardening renders channel estimation at the terminals, and the associated transmission of downlink pilots, unnecessary in most cases.

Another benefit of channel hardening in Massive MIMO is that the effective scalar channel seen by each terminal behaves much like an additive white Gaussian noise (AWGN) channel, and hence standard coding and modulation techniques devised for the AWGN channel tend to work well. To illustrate this point, consider the empirical link performance example shown in Figure 1.6. Here, an array with $M = 100$ antennas serves $K = 40$ terminals that transmit simultaneously in the uplink, using QPSK modulation and a rate-$1/2$ channel code,

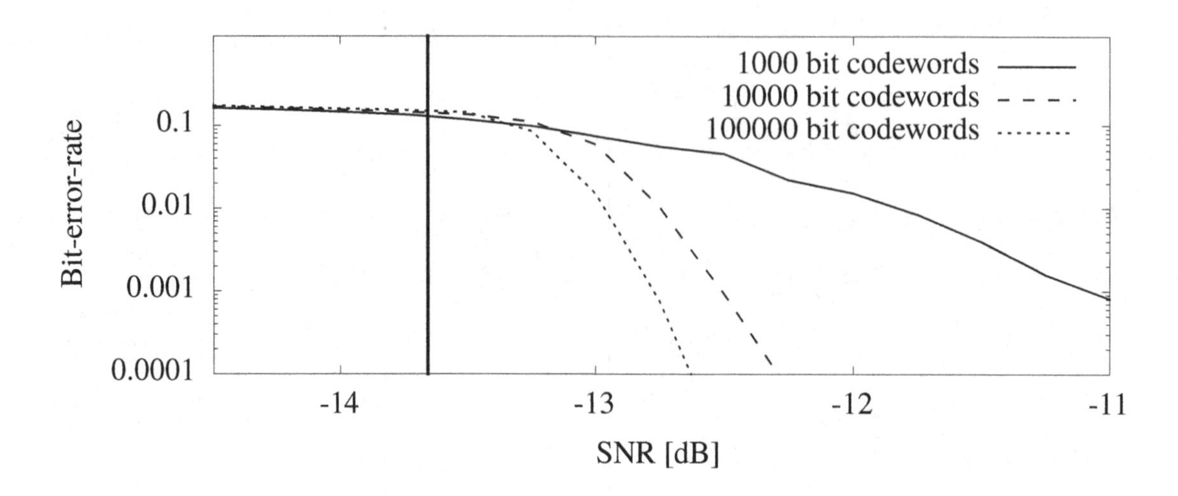

Figure 1.6. Empirical performance of a Massive MIMO uplink with $M = 100$ antennas and $K = 40$ terminals using QPSK modulation with a rate-$1/2$ low-density parity check code. The vertical solid line represents the SNR threshold obtained from the closed-form lower bound on the spectral efficiency derived in Chapter 3.

with a coherence interval (see Chapter 2 for the exact definition) length of 400 samples. The fading is Rayleigh and independent between the antennas and between the coherence intervals, and there is coding across coherence intervals. All terminals have the same path loss and transmit with the same power, and there is no shadow fading. The channel code is a state-of-the-art low-density parity check code optimized for the AWGN channel [27]. The base station learns the uplink channels through received pilot signals, and each terminal transmits its own orthogonal pilot sequence of length 40. The instantaneous sum spectral efficiency is equal to the number of terminals, K, times the number of bits per symbol, times the code rate: $40 \times 2 \times (1/2) = 40$ b/s/Hz. Assuming, for the sake of argument, that there is only transmission in the uplink, the net sum spectral efficiency is equal to the fraction of the coherence interval spent on payload data transmission multiplied by the instantaneous sum spectral efficiency: $(1 - 40/400) \times 40 = 36$ b/s/Hz. The receiver performs maximum-ratio processing followed by channel decoding. The lower bound on the instantaneous ergodic sum spectral efficiency for this case is $K \log_2(1 + \text{SINR})$ where SINR is the effective SINR given in the upper right corner of Table 3.1 (the same for all terminals). Equating this bound to 40 b/s/Hz, we find that the minimum required SNR is -13.66 dB. For a block length of 100 000 bits, the SNR gap to the bound is about 1 dB. While not shown in Figure 1.6, the smaller the ratio K/M, the closer is the effective channel seen by each terminal to an AWGN channel, and the smaller is the gap to the corresponding capacity bound.

The above example demonstrates both that off-the-shelf coding and modulation techniques tend to work well in Massive MIMO, and that the closed-form spectral-efficiency bounds derived in this book can be closely approached in practice. Hence, these bounds are eminently suitable as proxy for the link performance, when working with system optimization, resource allocation, and power control. Chapter 6 provides design examples based on these bounds.

1.4 Time-Division versus Frequency-Division Duplexing

Point-to-Point MIMO, conventional Multiuser MIMO, and Massive MIMO require different amounts of CSI at the base station and at the terminals. This CSI may be obtained either by estimation from received pilot signals, or by feedback from the receiver to the transmitter, or both:

- In *time-division duplexing* (TDD) operation, the base station learns the uplink channel from uplink pilots. In addition, because the channel is reciprocal (the impulse response between any two antennas is the same in both directions) once the base station has learned the uplink channel, it automatically has a legitimate estimate of the downlink channel. Massive MIMO, in the form described above and throughout this book, assumes TDD operation.
- In *frequency-division duplexing* (FDD) operation, the terminals learn the downlink channel from pilots sent by the base station, and communicate the estimated CSI back to the base station over a control channel. This feedback can be very costly, except in special cases, such as in LoS propagation, when the CSI can be efficiently quantized. If the CSI possesses no special structure, then direct analog feedback may be as efficient as any digital scheme [28].

  To learn the uplink channel, the base station listens to pilots sent by the terminals.

Learning the channel by sending pilots consumes resources. More specifically, to facilitate channel estimation at the receiver, during each segment of the time-frequency plane over which the channel is static (called a *coherence interval* and defined more precisely in Chapter 2), each transmitting antenna needs to be assigned a unique pilot waveform, and all such pilots need to be mutually orthogonal. This means that, for example, in FDD, if M antennas transmit orthogonal pilots in the downlink, then at least M samples per coherence interval have to be spent on pilots.

Table 1.1 summarizes the amount of resources needed for pilots and CSI feedback for Point-to-Point MIMO, Multiuser MIMO, and Massive MIMO. Massive MIMO operating in TDD mode stands out, because the amount of pilot resources required is independent of the number of base station antennas, M. Moreover, feedback from the terminals is avoided entirely. Consequently, Massive MIMO with TDD operation has unlimited scalability in M, a central motivation of the Massive MIMO concept.

As we shall see, even Massive MIMO operating in TDD has ultimate limitations. In particular, when terminals have mobility there is only time for the creation of a limited number of orthogonal pilots. In multi-cell systems, pilots have to be reused to some degree from cell to cell, which contaminates channel estimates in the home cell with channels from other cells. This phenomenon, called *pilot contamination*, results in degradation of the channel estimate quality and coherent interference that does not disappear with the addition of more antennas.

In one respect, FDD offers an advantage over TDD. Specifically, under a peak power constraint and noise-limited operation, FDD yields 3 dB better SNR than TDD. More exactly, let B be the total system bandwidth for both uplink and downlink, let P be the received power at the terminal during the time that the transmitter is active, and let N_0 be the noise spectral density. With TDD, the net downlink rate is

$$\frac{B}{2}\log_2\left(1+\frac{P}{BN_0}\right) \tag{1.5}$$

b/s, where the division of B by 2 before the logarithm reflects the fact that transmission takes place over the full bandwidth, but only half of the time. With FDD, the net rate is

$$\frac{B}{2}\log_2\left(1+\frac{P}{(B/2)N_0}\right) \tag{1.6}$$

b/s, where B is divided by 2 both inside and outside of the logarithm because transmission takes place continuously, but only over half the bandwidth. The discrepancy between (1.5) and (1.6) is solely a consequence of the fact that with TDD the transmitter is silent half of the time, so for a given P the received energy per unit time is only half that of FDD. For mobile terminals, the greater multiplexing capability of TDD compared with FDD will easily offset this 3 dB loss in coverage. Moreover, under interference-limited operation, the 3 dB gap tends to disappear because the interference power is the same for TDD and FDD.

	FDD		TDD	
	Uplink	Downlink	Uplink	Downlink
Point-to-Point MIMO (no CSI at the transmitter)	K pilots	M pilots	K pilots	M pilots
Conventional Multiuser MIMO	K pilots + M CSI coeff.	M pilots	K pilots	M pilots
Massive MIMO	K pilots + M CSI coeff.	M pilots	K pilots	none

Table 1.1. Minimum possible resources consumed by pilot transmission and CSI feedback in terms of total number of samples per coherence interval for the three variants of MIMO. CSI feedback assumes that $M \geq K$, and analog feedback.

1.5 Summary of Key Points

- With Point-to-Point MIMO, see Figure 1.1, terminals are orthogonally multiplexed. Both the base station and the terminals have multiple antennas. Point-to-Point MIMO is unscalable because of the required pilot overhead, and because LoS propagation yields channels of insufficient rank. More detail on Point-to-Point MIMO is given in Section C.3.

- With Multiuser MIMO, see Figure 1.3, terminals are spatially multiplexed. Each terminal may have a single antenna, as assumed throughout this book. The requirements on the propagation channel are substantially relaxed as compared to Point-to-Point MIMO. However, optimal signal processing is complicated, and accurate two-way CSI is required, which in turn demands that large resources be devoted to pilots. More detail on Multiuser MIMO is given in Section C.4.

- Massive MIMO, the ultimate form of Multiuser MIMO (see Figure 1.4), stands out from conventional Multiuser MIMO in several ways. First, only the base station obtains CSI. Thanks to channel hardening, no channel estimation is required at the terminals. By operating in TDD mode and exploiting reciprocity of the propagation channel, the amount of resources needed for pilots only depends on the number of simultaneously served terminals, K. This renders Massive MIMO entirely scalable with respect to the number of base station antennas, M. When M is large, linear processing at the base station is nearly optimal.

Chapter 2

MODELS AND PRELIMINARIES

This chapter introduces the basic signal and channel models to be used throughout the book. We use standard complex baseband representations of all signals and noise, with the implicit assumption that all signals are eventually modulated onto a carrier with frequency f_c and wavelength $\lambda = c/f_c$, where c is the speed of light. Also, unless stated explicitly, all Gaussian random variables are complex-valued and circularly symmetric; see Appendix A for a treatment of such variables.

2.1 Single-Antenna Transmitter and Single-Antenna Receiver

The wireless channel takes an input signal $x(t)$, emitted by a transmit antenna, and yields an output signal $y(t)$, observed at a receive antenna. The relation between $x(t)$ and $y(t)$ is linear, owing to the linearity of Maxwell's equations. However, this relation generally is time-varying, since the transmitter, receiver, and other objects in the propagation environment may move relative to one another.

2.1.1 Coherence Time

The time during which the channel can be reasonably well viewed as time-invariant is called the *coherence time* and denoted by T_c (measured in seconds). To relate T_c to the characteristics of the physical propagation environment, we consider a simple two-path propagation model where a transmit antenna emits a signal $x(t)$ that reaches the receiver both directly via a LoS path, and via a single specular reflection; see Figure 2.1(a). If both paths have unit strength, and the bandwidth of $x(t)$ is small enough that time-delays can be

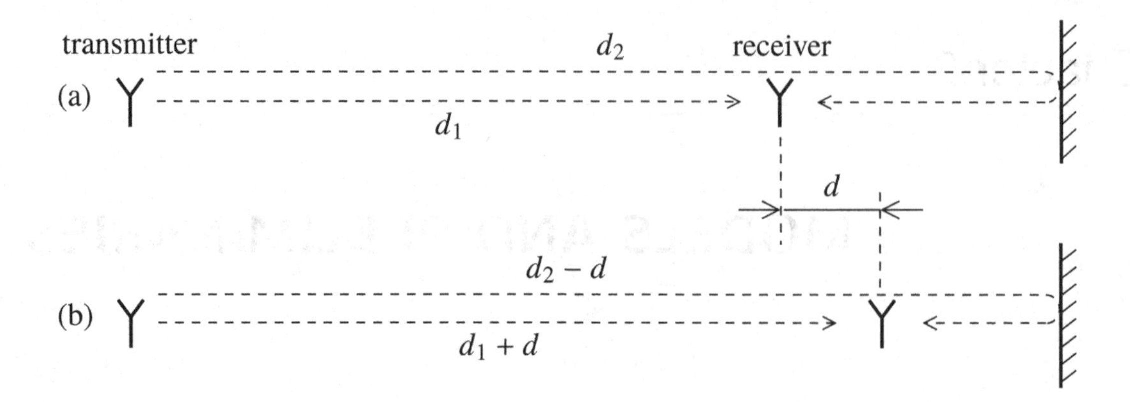

Figure 2.1. Illustration of the two-path propagation model used to motivate the definitions of coherence time and coherence bandwidth.

approximated as phase shifts, then by the superposition principle the received signal is

$$\begin{aligned} y(t) &= \left(e^{-i2\pi f_c \frac{d_1}{c}} + e^{-i2\pi f_c \frac{d_2}{c}}\right) x(t) \\ &= \left(e^{-i2\pi \frac{d_1}{\lambda}} + e^{-i2\pi \frac{d_2}{\lambda}}\right) x(t), \end{aligned} \tag{2.1}$$

where d_1 and d_2 are the propagation path lengths defined in Figure 2.1(a).

Suppose, for the sake of argument, that when the receiver is located as shown in Figure 2.1(a), d_1/λ and d_2/λ are integers. Then the two paths add up constructively and $y(t) = 2x(t)$. Next, if the receiver is displaced d meters to the right, so that we have the situation in Figure 2.1(b), the received signal will instead be

$$\begin{aligned} y(t) &= \left(e^{-i2\pi \frac{d}{\lambda}} + e^{-i2\pi \frac{-d}{\lambda}}\right) x(t) \\ &= 2\cos\left(2\pi \frac{d}{\lambda}\right) x(t). \end{aligned} \tag{2.2}$$

The two paths add up destructively if the cosine in (2.2) is equal to zero. As shown in Figure 2.2(a), this occurs periodically for displacements d that are spaced $\lambda/2$ meters apart. The channel may be considered time-invariant as long as the receiver does not move farther than this distance, $\lambda/2$. This means that if the receiver moves with velocity v meters/second, then the coherence time, T_c, is

$$T_c = \frac{\lambda}{2v} \quad \text{seconds.} \tag{2.3}$$

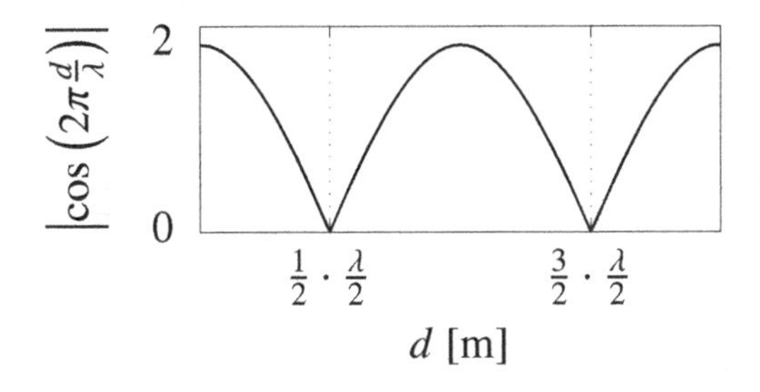

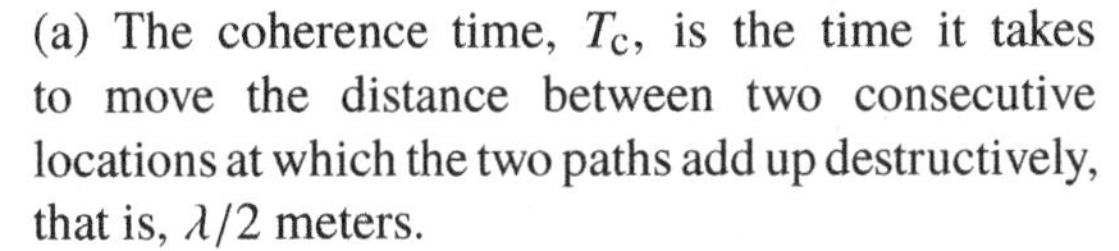

(a) The coherence time, T_c, is the time it takes to move the distance between two consecutive locations at which the two paths add up destructively, that is, $\lambda/2$ meters.

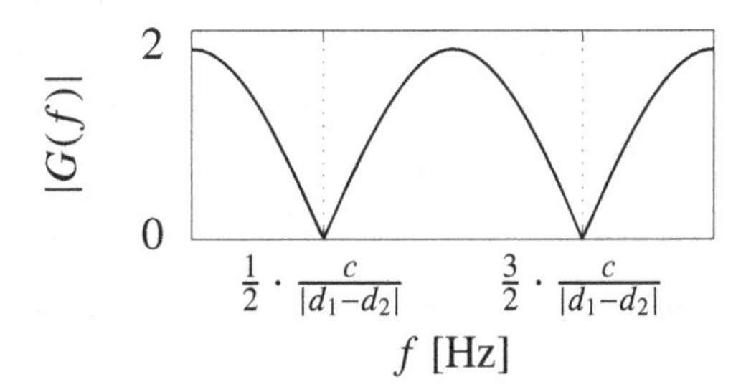

(b) The coherence bandwidth, B_c, is the frequency separation between two nulls of the frequency response $G(f)$, that is, $c/|d_1 - d_2|$ Hz.

Figure 2.2. Definitions of coherence time and coherence bandwidth for the two-path model in Figure 2.1.

A real propagation environment is considerably more involved than the two-path model of Figure 2.1. It can entail a direct path and a multiplicity of indirect paths via scattering centers of different amplitudes. The overall response is generally complex-valued. Nevertheless, the coherence time as specified by (2.3) is typically a good approximation.

2.1.2 Coherence Bandwidth

Consider now the transmission of a waveform whose time-duration is shorter than the coherence time, T_c. The relation between $x(t)$ and $y(t)$ is then approximately time-invariant, and defined by the channel impulse response $g(t)$ (where $y(t) = \int_{-\infty}^{\infty} d\tau\, g(\tau)x(t-\tau)$) or, equivalently, by the channel frequency response

$$G(f) = \int_{-\infty}^{\infty} dt\, g(t)e^{-i2\pi ft}. \tag{2.4}$$

Generally, the magnitude of the channel frequency response, $|G(f)|$, varies with f. The length of a frequency interval over which $|G(f)|$ is approximately constant is called the *coherence bandwidth* and denoted by B_c (measured in Hz). Consider again the two-path propagation model in Figure 2.1(a), and assume that d_1 and d_2 are fixed and chosen such that d_1/λ and d_2/λ are integers. If a sinusoidal signal, $x(t) = e^{i2\pi ft}$, is transmitted, then the received signal is

$$y(t) = \left(e^{-i2\pi(f_c+f)\frac{d_1}{c}} + e^{-i2\pi(f_c+f)\frac{d_2}{c}}\right)e^{i2\pi ft}. \tag{2.5}$$

Hence, the frequency response of the channel is

$$G(f) = e^{-i2\pi(f_c+f)\frac{d_1}{c}} + e^{-i2\pi(f_c+f)\frac{d_2}{c}}$$
$$= e^{-i2\pi f\frac{d_1}{c}} + e^{-i2\pi f\frac{d_2}{c}}. \tag{2.6}$$

The magnitude of the frequency response is

$$|G(f)| = \left|e^{-i2\pi f\frac{d_1}{c}} + e^{-i2\pi f\frac{d_2}{c}}\right|$$
$$= 2\left|\cos\left(\pi f\frac{d_1 - d_2}{c}\right)\right|, \tag{2.7}$$

independently of f_c. $|G(f)|$ has zero-crossings at frequencies periodically spaced $c/|d_1 - d_2|$ Hz apart; see Figure 2.2(b). Analogously to the definition of coherence time in (2.3), we define the coherence bandwidth B_c to be the spacing between two nulls of $|G(f)|$, that is

$$B_c = \frac{c}{|d_1 - d_2|} \quad \text{Hz}. \tag{2.8}$$

While the two-path model represents a simplified description of reality, in practice we expect $|G(f)|$ to be substantially constant over a frequency interval whose length is given by (2.8), where $|d_1 - d_2|$ is the maximum difference in length between different propagation paths from the transmitter to the receiver. As a first-order approximation, $|d_1 - d_2|/c$ is equal to the delay spread of the channel, and $g(t)$ is time-limited to $|d_1 - d_2|/c$ seconds.

2.1.3 Coherence Interval

A time-frequency space of duration T_c seconds and bandwidth B_c Hz is called a *coherence interval*. This is the largest possible time-frequency space within which the effect of the channel reduces to a multiplication by a complex-valued scalar gain g. The magnitude $|g|$ represents the scaling of the waveform envelope and $\arg(g)$ represents the shift in its phase.

According to the sampling theorem, any T-second segment of a waveform $x(t)$ whose energy is substantially contained in a B Hz wide frequency interval can be described in terms of BT (complex-valued) samples taken at intervals of $1/B$ seconds. This means that B_cT_c (complex-valued) samples are required to define a waveform that fits into one coherence interval. We therefore say that a coherence interval has the length

$$\tau_c = B_cT_c \quad \text{samples}. \tag{2.9}$$

	Indoors $\lvert d_1 - d_2 \rvert = 30$ meters	Outdoors $\lvert d_1 - d_2 \rvert = 1\,000$ meters
Pedestrian $v = 1.5$ m/s (5.4 km/h)	$B_c = 10$ MHz $T_c = 50$ ms $\tau_c = 500\,000$	$B_c = 300$ kHz $T_c = 50$ ms $\tau_c = 15\,000$
Vehicular $v = 30$ m/s (108 km/h)	N/A	$B_c = 300$ kHz $T_c = 2.5$ ms $\tau_c = 750$

Table 2.1. First-order estimates of the coherence time T_c, coherence bandwidth B_c, and sample length of the coherence interval, τ_c, for some different propagation scenarios, at a carrier frequency of 2 GHz (λ =15 cm).

Consider a waveform $x(t)$, occupying one coherence interval, transmitted over a channel having the same coherence interval. The output of the noisy channel, sampled at rate B_c, takes the form

$$y_n = gx_n + w_n, \quad n = 0, \ldots, \tau_c - 1, \tag{2.10}$$

where x_n is the input, y_n is the output, g represents the channel gain, and $\{w_n\}$ denote samples of additive receiver noise. Throughout the book, we assume that the noise is a stationary random process having flat bandlimited spectral support, $[-B_c, B_c]$. The noise autocorrelation function is proportional to $\text{sinc}(B_c t)$, so noise samples $\{w_n\}$ taken at intervals of $1/B_c$ seconds are uncorrelated.

Some typical values of B_c, T_c, and τ_c, computed using (2.3), (2.8), and (2.9), are shown in Table 2.1 for a carrier frequency of $f_c = 2$ GHz. The values of $\lvert d_1 - d_2 \rvert$ depend on the exact characteristics of the propagation environment, and the values in the table are only first-order estimates. An important observation is that outdoors in high mobility, τ_c is only on the order of a few hundred samples.

2.1.4 Interpretation of T_c and B_c in Terms of Nyquist Sampling Rate

It will be convenient in our subsequent analyses to pretend that the channel is static during each coherence interval, as in (2.10). In reality, however, the amplitude and phase of g smoothly evolve between consecutive samples and therefore have some dependence on n. In a practical system, estimates of g acquired from pilots may require interpolation for which the sampling theorem provides a rational basis.

The sampling theorem applies rigorously to a function that is strictly bandlimited to $[-B, B]$ and is sampled at intervals of $1/B$ for all time. Our definitions of coherence time and coherence bandwidth are equivalent to specifications of Nyquist sampling intervals for functions that are substantially bandlimited. Thus, coherence time (2.3) is associated with motion over half of one cycle of a sinewave, while coherence bandwidth (2.8) is equivalent to the reciprocal of the channel delay spread. In practice, one may not be dealing with strictly bandlimited functions (in particular, if there is strong near-field propagation, or reverberation), but there is still ample precedent for invoking the sampling theorem. The concept of the coherence interval (2.9), while exceedingly useful, is somewhat nebulous, and in actual systems the nominal interval may have to be shortened to provide an adequate design margin, especially in case a terminal is served over only a few consecutive coherence intervals, or if residual carrier frequency offsets remain, or to accommodate special applications that require high-accuracy interpolation. In the case studies of Chapter 6, we adopt a factor of two design margin in specifying the coherence interval.

2.1.5 TDD Coherence Interval Structure

As pointed out in Section 1.4, TDD operation is ideal for Massive MIMO because the training burden is independent of the number of base station antennas. Throughout the book, we assume half-duplex TDD so that only one end of the link is transmitting at any one time, either the base station or the terminals. As a consequence, the coherence interval naturally divides into uplink and downlink subintervals, not necessarily of equal duration. Figure 2.3 illustrates two possible configuration, where Figure 2.3(a) includes provision for downlink as well as uplink pilots, and Figure 2.3(b) has only uplink pilots. Not shown are guard intervals between uplink and downlink transmissions.

Let τ_{ul} be the number of samples per coherence interval spent on transmission of uplink payload data, $\tau_{\mathrm{ul,p}}$ the number of samples per coherence interval spent on uplink pilots, τ_{dl} the number of samples used for transmission of downlink payload data, and $\tau_{\mathrm{dl,p}}$ the number of samples allocated for downlink pilots. For the Figure 2.3(a) structure,

$$\tau_{\mathrm{ul}} + \tau_{\mathrm{ul,p}} + \tau_{\mathrm{dl,p}} + \tau_{\mathrm{dl}} = \tau_{\mathrm{c}}. \tag{2.11}$$

We show later that uplink pilots alone are sufficient to make TDD Massive MIMO work, and for the remainder of the book we assume the coherence interval structure of Figure 2.3(b). For the sake of simplicity, we drop the subscript $(\cdot)_{\mathrm{ul}}$ from the parameter $\tau_{\mathrm{ul,p}}$, and the structural constraint becomes

$$\tau_{\mathrm{ul}} + \tau_{\mathrm{p}} + \tau_{\mathrm{dl}} = \tau_{\mathrm{c}}. \tag{2.12}$$

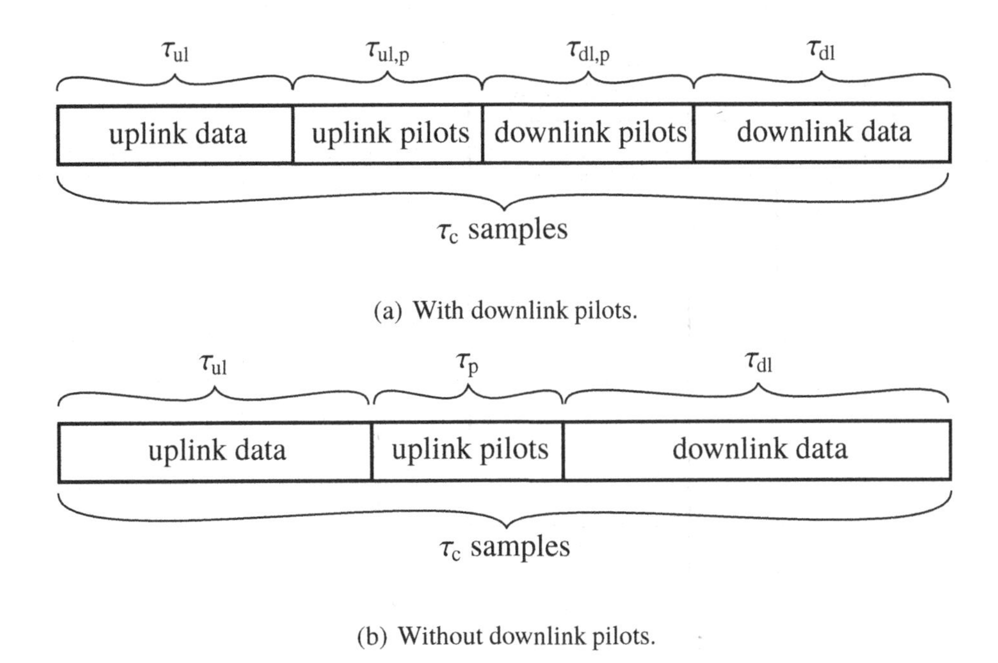

Figure 2.3. Allocation of the samples in a coherence interval.

2.1.6 The Coherence Interval in the Context of OFDM Modulation

Orthogonal frequency-division multiplexing (OFDM) is a popular modulation scheme that is fundamentally simple and has numerous attractive properties. The use of OFDM also facilitates a natural interpretation of the coherence interval concept. However, nothing said in this book is specific to OFDM. All spectral efficiency results to be given in subsequent chapters are valid regardless of the particular modulation scheme that is eventually used in an implementation.

OFDM uses the (fast) discrete Fourier transform to decompose a frequency-selective channel into many parallel channels called *subcarriers*; see Figure 2.4. By virtue of the *cyclic prefix* (see below), the effect of the channel on each subcarrier is purely multiplicative, and each subcarrier sees a flat-fading channel. While there are other versions of OFDM, here we treat only the variant that uses a cyclic prefix.

Transmission entails a sequence of OFDM symbols, each symbol consisting of a useful part of length T_u seconds, preceded by a cyclic prefix (also known as guard interval) of T_{cp} seconds. In total, each OFDM symbol is $T_s = T_{cp} + T_u$ seconds long; see Figure 2.5. The useful part carries N_s samples which are obtained by discrete Fourier-transformation of N_s information symbols, and the cyclic prefix replicates the T_{cp} last seconds of the useful part.

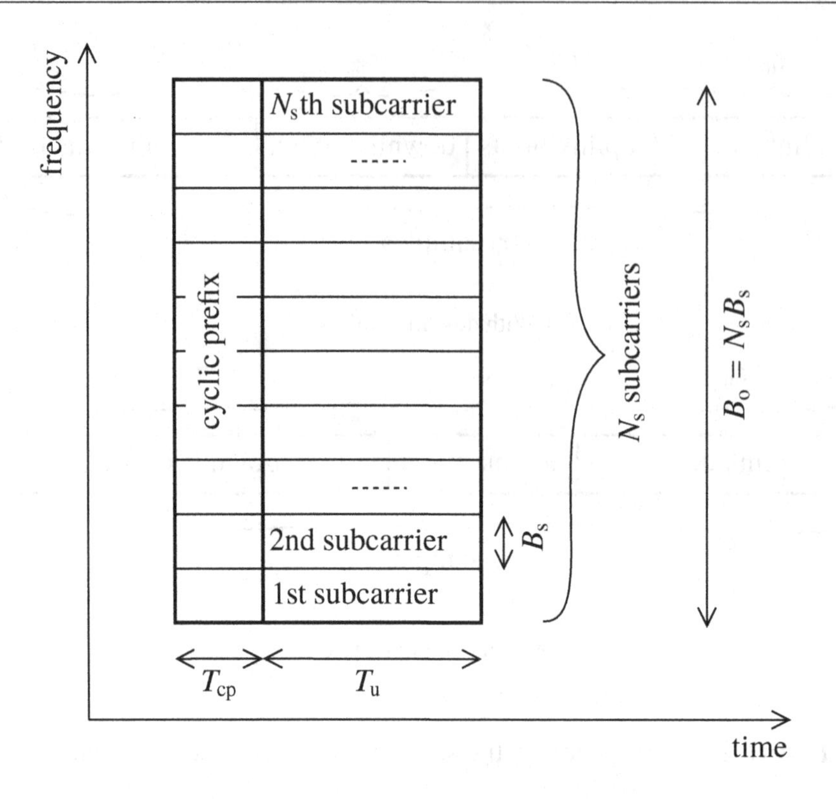

Figure 2.4. Time-frequency domain view of an OFDM symbol, including its cyclic prefix.

The effect of prepending the cyclic prefix is that the linear convolution that represents the effect of the channel impulse response is transformed into a circular convolution, which is equivalent to multiplication in the frequency domain. OFDM renders the original wideband delay-spread channel into many parallel narrowband flat-fading channels. The number of subcarriers is N_s and the frequency separation between neighboring subcarriers is $B_s = 1/T_u$. Hence, the total bandwidth occupied by an OFDM symbol is

$$B_o = N_s B_s = \frac{N_s}{T_u}. \tag{2.13}$$

In order for consecutive OFDM symbols not to interfere, T_{cp} must be at least as large as the channel delay spread.

In practice, several consecutive OFDM symbols are grouped together into one *slot*. We denote the number of OFDM symbols in a slot by N_{slot}, and the duration of one slot by

$$T_{slot} = N_{slot} T_s. \tag{2.14}$$

We assume that $T_{slot} \le T_c$, so that the channel is time-invariant during one slot. However,

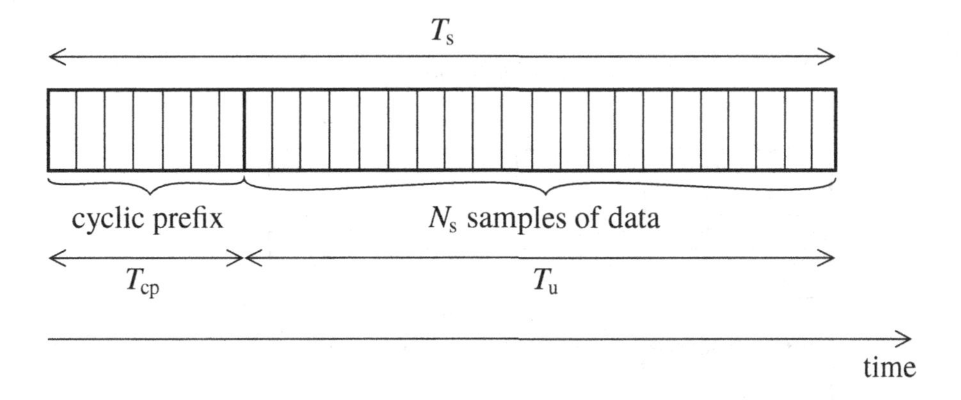

Figure 2.5. Structure of an OFDM symbol, in the time domain.

T_{slot} is not necessarily equal to T_c. In fact, it may be expedient to use a slot that is shorter than the coherence time, say if reduced latency were a consideration.

Normally, the total OFDM symbol bandwidth, B_o, is much greater than the channel coherence bandwidth, B_c, while the subcarrier bandwidth B_s is smaller than B_c. We denote the number of consecutive subcarriers in the frequency domain that fit into one coherence bandwidth by N_{smooth}, assumed to be an integer here. Then,

$$B_c = N_{smooth} B_s. \tag{2.15}$$

The number N_{smooth} represents the number of subcarriers over which the channel frequency response is smooth (approximately constant).

If $T_c = T_{slot}$, a coherence interval consists of N_{smooth} neighboring subcarriers in the frequency domain and N_{slot} consecutive OFDM symbols in the time domain. Each slot consists of

$$\frac{N_s}{N_{smooth}} = \frac{B_o}{B_c} \tag{2.16}$$

coherence intervals that are located adjacent to one another in the frequency domain; see Figure 2.6. The length of a coherence interval, measured in samples is

$$\begin{aligned} B_c T_c &= B_c T_{slot} \\ &= N_{smooth} B_s N_{slot} T_s \\ &= \frac{N_{smooth}}{T_u} N_{slot} T_s \\ &= \frac{T_s}{T_u} N_{smooth} N_{slot}. \end{aligned} \tag{2.17}$$

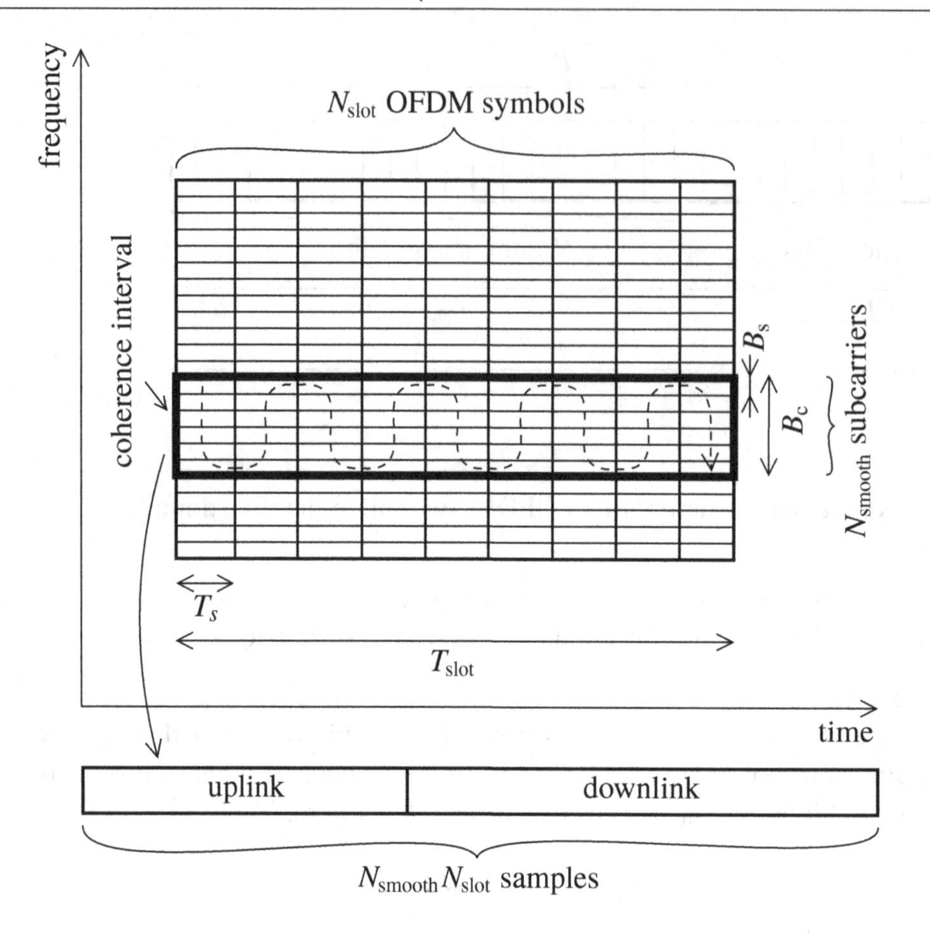

Figure 2.6. A slot comprises N_{slot} consecutive OFDM symbols, each of which contains N_{s} subcarriers. If $T_{\text{c}} = T_{\text{slot}}$, a coherence interval spans N_{slot} OFDM symbols and N_{smooth} subcarriers and each slot contains $N_{\text{s}}/N_{\text{smooth}}$ coherence intervals. The lower part of the figure shows a possible mapping between the time-frequency domain and the samples in a coherence interval.

A fraction $T_{\text{u}}/T_{\text{s}}$ of each coherence interval is useful, and the rest is spent on the cyclic prefix. Thus, the number of useful samples per coherence interval is

$$\frac{T_{\text{u}}}{T_{\text{s}}} B_{\text{c}} T_{\text{c}} = N_{\text{smooth}} N_{\text{slot}}, \tag{2.18}$$

which is equal to τ_{c} as defined in (2.9), up to a factor that reflects the loss of useful samples associated with the cyclic prefix. All samples in a coherence interval are affected by a scaling with a channel gain g as in (2.10). Figure 2.6 also shows a possible mapping between the time-frequency domain and the $N_{\text{smooth}} N_{\text{slot}}$ samples in a coherence interval.

Table 2.2 shows parameters of a sample OFDM system, where the channel delay spread is

OFDM symbol duration	T_s	$\frac{1}{14}$ ms
OFDM symbol duration, useful part	T_u	$\frac{1}{15}$ ms
Cyclic prefix duration	T_{cp}	$\frac{1}{14\times 15}$ ms
Subcarrier spacing	B_s	15 kHz
Coherence bandwidth	B_c	210 kHz
Number of subcarriers within coherence bandwidth	N_{smooth}	14
Slot duration	T_{slot}	2 ms
Number of OFDM symbols within one slot	N_{slot}	28
Number of useful samples per coherence interval, if $T_c = T_{slot}$	$N_{smooth}N_{slot}$	392

Table 2.2. Parameters of a sample OFDM system.

assumed equal to the duration of the cyclic prefix, T_{cp}.

2.1.7 Small-Scale and Large-Scale Fading

Within a coherence interval, as illustrated in Figure 2.6 (for OFDM), the complex-valued gain between any pair of antennas is substantially constant, and is denoted by the symbol g. It is useful to factor g as follows:

$$g = \sqrt{\beta}h. \tag{2.19}$$

The positive real number, β, called the *large-scale fading* coefficient, embodies range-dependent path loss and shadow fading, it is virtually independent of frequency, and is strongly correlated over many wavelengths of space. The complex-valued number h, representing *small-scale fading*, models range dependent phase shift and constructive and destructive interference among different propagation paths.

In all ensuing analyses, we will assume that the small-scale fading is Rayleigh; that is, $h \sim \mathcal{CN}(0, 1)$. The assumption of Rayleigh fading permits the use of Bayesian analysis and it makes ergodic capacity a legitimate performance criterion. Rayleigh fading is also straightforward to justify with simple physical models. For example, in isotropic scattering, h represents the combined effect of many independent propagation paths so by the superposition principle and the central limit theorem, h will be approximately circularly symmetric Gaussian. While all quantitative performance analyses in this book rest on the Rayleigh fading assumption, in Chapter 7 we will argue that under the extreme opposite

LoS propagation regime, Massive MIMO is still fully functional. We will assume that large-scale fading coefficients are known a priori to anyone who needs to know them, but that small-scale fading is a priori known to nobody.

2.1.8 Normalized Signal Model and SNR

Henceforth, we will work with the following normalized model for the received signal:

$$y = \sqrt{\rho} g x + w, \tag{2.20}$$

where w is receiver noise and ρ is a dimensionless constant that scales the transmitted signal. Throughout, we adopt the convention that each transmitted signal x has zero mean and satisfies a unit power constraint, $\mathsf{E}\{x\} = 0$ and $\mathsf{E}\left\{|x|^2\right\} \leq 1$. We also assume that the noise is circularly symmetric Gaussian with unit variance, denoted $w \sim \mathcal{CN}(0, 1)$, and is independent of x. This gives ρ the interpretation of a *signal-to-noise ratio* (SNR) in the following sense: if the median of β equals unity, and the transmitter expends its maximum permitted power, then ρ is the median SNR measured at the receiver.

We denote the SNR associated with the uplink and downlink by ρ_{ul} respectively ρ_{dl}. Hence, on the uplink,

$$y = \sqrt{\rho_{\text{ul}}} g x + w, \tag{2.21}$$

and on the downlink,

$$y = \sqrt{\rho_{\text{dl}}} g x + w, \tag{2.22}$$

where in both cases, x is the transmitted signal, y is the received signal, and w represents noise. The uplink and downlink SNRs are different in general, owing to differences in the transmit powers and the noise figures at the base station and the terminal.

2.2 Multiple Base Station Antennas and Multiple Terminals

We now consider cellular Massive MIMO whereby base stations, each equipped with an array of M antennas, serve simultaneously a multiplicity of terminals in their designated areas via spatial multiplexing. We introduce the signal models and assumptions that underlie the spectral efficiency analysis in the ensuing chapters.

We confine the discussion entirely to the case when the terminals have a single antenna each. The case of multiple-antenna terminals can be handled, for example, by treating each terminal antenna as a separate terminal.

2.2.1 Single-Cell System

We first consider the case of a single base station that simultaneously serves K terminals. We call the area where the terminals are located a *cell*, and refer to the corresponding scenario as *single-cell*, emphasizing the fact that there is no inter-cell interference to account for.

Let g_k^m be the channel gain between the kth terminal and the mth base station antenna; see Figure 2.7. We will assume that the base station antennas are configured in a compact array, so that the paths between a given terminal and all base station antennas are affected by the same large-scale fading, but by different small-scale fading. Hence,

$$g_k^m = \sqrt{\beta_k} h_k^m, \qquad k = 1, \ldots, K, \qquad m = 1, \ldots, M, \tag{2.23}$$

where β_k is a large-scale fading coefficient that depends on k but not on m, and h_k^m represents the effect of small-scale fading. We let $\boldsymbol{G}$ be a matrix that comprises the channel gains between all terminals and all base station antennas,

$$\boldsymbol{G} = \begin{bmatrix} g_1^1 & \cdots & g_K^1 \\ \vdots & \ddots & \vdots \\ g_1^M & \cdots & g_K^M \end{bmatrix}. \tag{2.24}$$

Throughout all performance analyses, we will assume that the small-scale fading is Rayleigh and *independent* between the antennas and the terminals, so that $\{h_k^m\}$ are i.i.d. $\mathrm{CN}(0, 1)$ random variables. We discuss this assumption further in Chapter 7.

Uplink

Consider the uplink. If the terminals simultaneously transmit the K signals $x_1, \ldots, x_K$, then the mth base station antenna receives the signal,

$$y_m = \sqrt{\rho_{\mathrm{ul}}} \sum_{k=1}^{K} g_k^m x_k + w_m, \tag{2.25}$$

where w_m is receiver noise. As before, we assume that $w_m \sim \mathrm{CN}(0, 1)$. Additionally, we will assume that the noise is uncorrelated across the antennas; that is, $\{w_m\}$ are independent. The transmit powers of the terminals are individually constrained,

$$\mathsf{E}\left\{|x_k|^2\right\} \le 1, \tag{2.26}$$

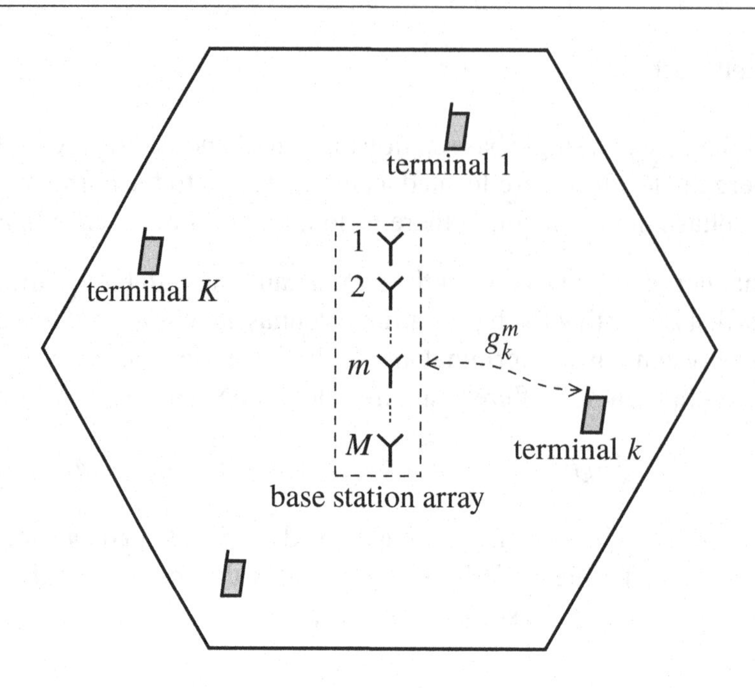

Figure 2.7. Single-cell propagation model.

as in Section 2.1.8, and the transmitted signals have zero mean: $\mathsf{E}\{x_k\} = 0$. Collectively, according to (2.25) the M antennas receive a vector $\boldsymbol{y} = [y_1, \ldots, y_M]^{\mathrm{T}}$,

$$\begin{aligned} \boldsymbol{y} &= \sqrt{\rho_{\mathrm{ul}}} \sum_{k=1}^{K} \boldsymbol{g}_k x_k + \boldsymbol{w} \\ &= \sqrt{\rho_{\mathrm{ul}}} \boldsymbol{G} \boldsymbol{x} + \boldsymbol{w}, \end{aligned} \tag{2.27}$$

where $\boldsymbol{g}_k$ is the kth column of $\boldsymbol{G}$, $\boldsymbol{x} = [x_1, \ldots, x_K]^{\mathrm{T}}$ and $\boldsymbol{w} = [w_1, \ldots, w_M]^{\mathrm{T}}$.

Downlink

On the downlink, the M base station antennas transmit the M-vector $\boldsymbol{x}$, and via reciprocity, the kth terminal receives

$$y_k = \sqrt{\rho_{\mathrm{dl}}} \boldsymbol{g}_k^{\mathrm{T}} \boldsymbol{x} + w_k, \tag{2.28}$$

where w_k is noise. In vector form,

$$\boldsymbol{y} = \sqrt{\rho_{\mathrm{dl}}} \boldsymbol{G}^{\mathrm{T}} \boldsymbol{x} + \boldsymbol{w}, \tag{2.29}$$

where $\boldsymbol{y} = [y_1, \ldots, y_K]^{\mathrm{T}}$ and $\boldsymbol{w} \triangleq [w_1, \ldots, w_K]^{\mathrm{T}}$. As before, we assume that the noise samples $\{w_k\}$ are i.i.d. CN(0, 1). Analogously to the single-antenna case in Section 2.1.8, we assume that $\boldsymbol{x}$ is normalized such that

$$\mathsf{E}\left\{\|\boldsymbol{x}\|^2\right\} \leq 1. \tag{2.30}$$

This normalization corresponds to enforcing a long-term constraint on the sum of the radiated power from all antennas. While this assumption is analytically convenient, it is not the only possibility. For example, one could alternatively consider power constraints for each antenna individually.

Signal-to-Noise Ratio

As in the single-antenna case, the quantities ρ_{ul} and ρ_{dl} have interpretations in terms of SNR. In the current context, on the uplink, if the median of β_k is unity for a given terminal and the terminal transmits with its maximum permitted power, then ρ_{ul} is the median SNR for that terminal, measured at any of the base station antennas. On the downlink, if the total permitted power were radiated through only one transmit antenna, say the first one, such that $\mathsf{E}\left\{|x_1|^2\right\} = 1$ and $x_2 = \cdots = x_M = 0$, and if additionally the median of β_k were equal to unity, then the median SNR measured at the kth terminal would be ρ_{dl}.

2.2.2 Multi-Cell System

Next we consider a *multi-cell* scenario. Here multiple base stations coexist, with some geographical separation, and each base station serves terminals in its associated cell. The antennas at each base station work coherently together, but different base stations do not cooperate. Generally, the carrier frequency used in a particular cell is reused in other cells, and inter-cell interference then results.

Throughout, we assume that there are K terminals in each cell. This assumption is made only for simplicity and, in reality, there may, of course, be a varying number of terminals in each cell. We will also assume synchronized operation, such that, at any given point in time, either all base stations simultaneously transmit or all terminals simultaneously transmit. This assumption is not strictly necessary, and it does not necessarily result in optimal system performance, but it is convenient to make for purposes of analysis.

Uplink

Consider first the uplink. The signal received at the mth base station antenna in the lth cell, denoted by y_{lm}, is a superposition of signals transmitted from the K terminals in the

lth cell, and the $K(L-1)$ terminals in all interfering cells $l' = 1, \ldots, l-1, l+1, \ldots, L$. Mathematically,

$$y_{lm} = \sqrt{\rho_{\text{ul}}} \sum_{k=1}^{K} g_{lk}^{lm} x_{lk} + \sqrt{\rho_{\text{ul}}} \sum_{\substack{l'=1 \\ l' \neq l}}^{L} \sum_{k=1}^{K} g_{l'k}^{lm} x_{l'k} + w_{lm}, \tag{2.31}$$

where $x_{l'k}$ is the signal transmitted by the kth terminal in the l'th cell and $g_{l'k}^{lm}$ is the channel gain from the kth terminal in the l'th cell to the mth base station antenna in the lth cell; see Figure 2.8. The last term in (2.31), w_{lm}, represents additive receiver noise, which we assume is $\mathcal{CN}(0, 1)$ and independent among different m and l. In vector form,

$$\boldsymbol{y}_l = \sqrt{\rho_{\text{ul}}} \boldsymbol{G}_l^l \boldsymbol{x}_l + \sqrt{\rho_{\text{ul}}} \sum_{\substack{l'=1 \\ l' \neq l}}^{L} \boldsymbol{G}_{l'}^l \boldsymbol{x}_{l'} + \boldsymbol{w}_l, \tag{2.32}$$

where $\boldsymbol{y}_l = [y_{l1}, \ldots, y_{lM}]^{\text{T}}$, $\boldsymbol{w}_l = [w_{l1}, \ldots, w_{lM}]^{\text{T}}$,

$$\boldsymbol{G}_{l'}^l = \begin{bmatrix} g_{l'1}^{l1} & \cdots & g_{l'K}^{l1} \\ \vdots & \ddots & \vdots \\ g_{l'1}^{lM} & \cdots & g_{l'K}^{lM} \end{bmatrix} \tag{2.33}$$

is an $M \times K$ matrix that contains all channel gains from the terminals in the l'th cell to the base station array in the lth cell, and the K-vector $\boldsymbol{x}_{l'} = [x_{l'1}, \ldots, x_{l'K}]^{\text{T}}$ contains the signals transmitted by the terminals in the l'th cell. In (2.33), analogous to (2.23),

$$g_{l'k}^{lm} = \sqrt{\beta_{l'k}^l} h_{l'k}^{lm}, \tag{2.34}$$

where $\beta_{l'k}^l$ models the large-scale fading associated with propagation from the kth terminal in the l'th cell to the base station array in the lth cell, and $h_{l'k}^{lm}$ models small-scale fading.

Downlink

Next, consider the downlink. Under the assumption of reciprocity, the kth terminal in the lth cell receives

$$y_{lk} = \sqrt{\rho_{\text{dl}}} \boldsymbol{g}_{lk}^{l\text{T}} \boldsymbol{x}_l + \sqrt{\rho_{\text{dl}}} \sum_{\substack{l'=1 \\ l' \neq l}}^{L} \boldsymbol{g}_{lk}^{l'\text{T}} \boldsymbol{x}_{l'} + w_{lk}, \tag{2.35}$$

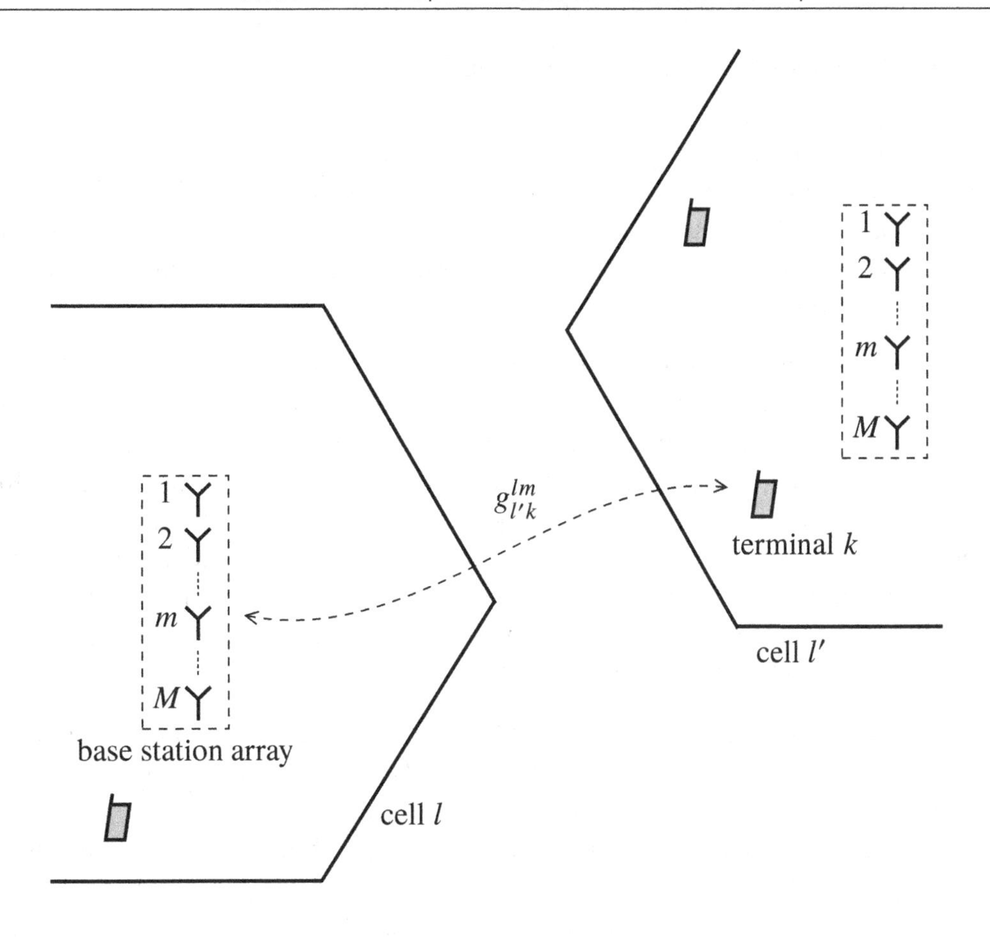

Figure 2.8. Multi-cell propagation model.

where $\boldsymbol{g}_{lk}^{l'}$ is the kth column of $\boldsymbol{G}_l^{l'}$, $\boldsymbol{x}_{l'}$ represents the M-vector transmitted by the array in the l'th cell, and w_{lk} is noise with distribution $\text{CN}(0, 1)$. Collectively, the K terminals in the lth cell will receive the K-vector,

$$\boldsymbol{y}_l = \sqrt{\rho_{\text{dl}}}\boldsymbol{G}_l^{l\text{T}}\boldsymbol{x}_l + \sqrt{\rho_{\text{dl}}}\sum_{\substack{l'=1\\l'\neq l}}^{L}\boldsymbol{G}_l^{l'\text{T}}\boldsymbol{x}_{l'} + \boldsymbol{w}_l, \tag{2.36}$$

where $\boldsymbol{y}_l = [y_{l1}, \ldots, y_{lK}]^{\text{T}}$ and $\boldsymbol{w}_l = [w_{l1}, \ldots, w_{lK}]^{\text{T}}$ is a K-vector of noise with i.i.d. $\text{CN}(0, 1)$ elements.

Similarly to the single-cell case, we will assume that the small-scale fading is Rayleigh and independent between all antennas and all terminals, so that $\{h_{l'k}^{lm}\}$ are i.i.d. $\text{CN}(0, 1)$ random variables.

2.3 Capacity Bounds as Performance Metric

In Massive MIMO, after appropriate signal processing, the effective channel associated with each of the terminals is a *scalar point-to-point channel.* Each time this channel is used, it takes a (complex-valued) scalar input symbol x and delivers a (complex-valued) output signal y. The action of the channel is characterized by the conditional probability distribution of y given x.

To communicate a message over a point-to-point scalar channel, the transmitter maps the message onto a sequence of symbols $\{x_n\}$, and the receiver recovers the message from the sequence of samples $\{y_n\}$. The effective number of bits conveyed per transmitted symbol, denoted by R, is called the *rate* and is measured in bits per channel use (bpcu). Recall that a waveform contained in a time-frequency space of bandwidth B Hz and time-duration T seconds can be described by BT samples; see Section 2.1.3. Hence, transmitting a waveform with bandwidth B Hz and time-duration T seconds is equivalent to transmitting BT symbols $\{x_n\}$. Therefore, the rate R is usually termed *spectral efficiency* and measured in bits per second per Hertz (b/s/Hz).

According to Shannon's noisy channel coding theorem, there exists a quantity C (unit: bpcu) called the channel *capacity*, which determines a rate R at which error-free communication is possible, asymptotically, when coding over many transmitted symbols. More precisely, the noisy channel coding theorem states that for any given probability of error ϵ, and any given "gap from capacity" ζ, there exist a block length N, and a coding scheme that achieves the rate $R = C - \zeta$ with a probability of a decoding error less than ϵ. Generally, achieving rates R that are close to C requires that N be large, and in the limit when ζ is forced towards zero, the required value of N tends to infinity.

For several channels, exact expressions for the capacity are known. In many cases, however, only bounds on capacity, also known as *achievable rates*, are available. Throughout this book, motivated by the channel coding theorem, we will use such capacity bounds as the primary performance metric. As illustrated in Section 1.3, these capacity bounds can typically be approached closely in practice by using state-of-the-art channel coding techniques.

In what follows, we present some key results on capacity and capacity bounds for point-to-point scalar channels that will be needed for the performance analysis in Chapters 3 and 4. Section C.2 contains comprehensive derivations of these results.

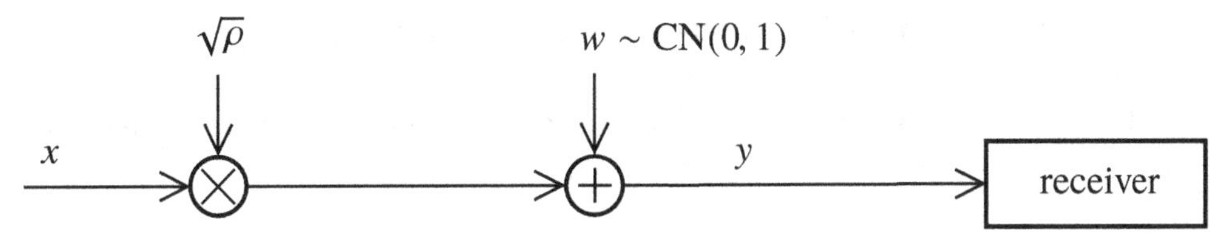

(a) Deterministic channel with additive Gaussian noise.

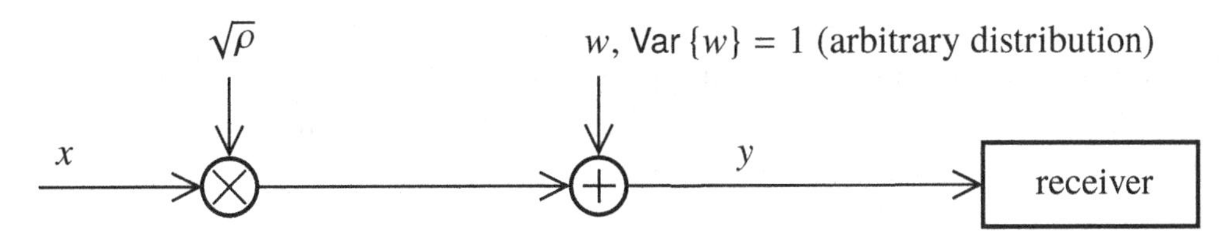

(b) Deterministic channel with additive non-Gaussian noise.

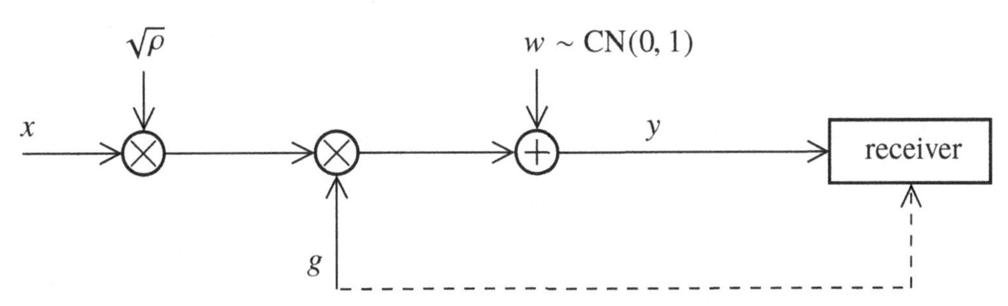

(c) Fading channel with additive Gaussian noise and perfect CSI at the receiver.

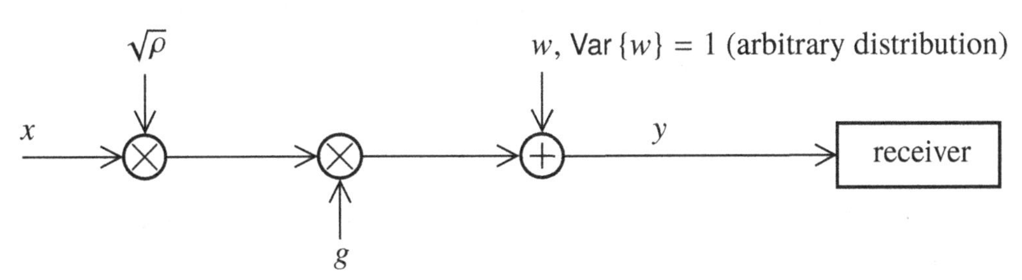

(d) Fading channel with additive non-Gaussian noise and no CSI at the receiver.

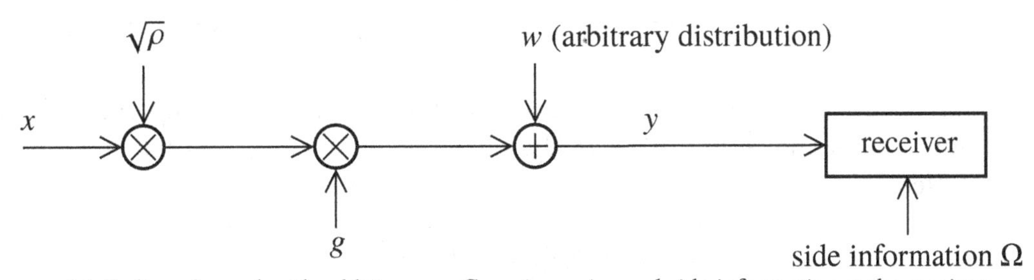

(e) Fading channel with additive non-Gaussian noise and side information at the receiver.

Figure 2.9. Scalar point-to-point channels.

2.3.1 Deterministic Channel with Additive Gaussian Noise

The most fundamental example of a scalar point-to-point channel is the deterministic channel with additive Gaussian noise; see Figure 2.9(a). Here,

$$y = \sqrt{\rho} x + w, \tag{2.37}$$

where w is noise that is independent of x and has distribution $\mathrm{CN}(0, 1)$, and ρ is a constant. A new independent realization of w is drawn for every transmitted symbol x. The transmitted symbol x satisfies the power constraint $\mathsf{E}\left\{|x|^2\right\} \leq 1$. Hence, ρ has the meaning of SNR, as in Section 2.1.8. The capacity of this channel is

$$C = \log_2 (1 + \rho), \tag{2.38}$$

and this capacity is achieved when the input symbols x are Gaussian distributed.

2.3.2 Deterministic Channel with Additive Non-Gaussian Noise

The next case of interest is when (2.37) applies, $\mathsf{E}\{w\} = 0$ and $\mathsf{Var}\{w\} = 1$, but w is not necessarily Gaussian; see Figure 2.9(b). Assuming that x and w are uncorrelated, but not necessarily independent,

$$\mathsf{E}\{x^* w\} = 0, \tag{2.39}$$

the capacity is lower bounded as follows:

$$C \geq \log_2 (1 + \rho). \tag{2.40}$$

In contrast to the Gaussian noise case in Section 2.3.1, the optimal distribution of the input symbol x is generally not Gaussian.

2.3.3 Fading Channel with Additive Gaussian Noise and Perfect CSI at the Receiver

We next introduce fading. The first model of interest is that of a fading channel with Gaussian noise and a gain g that is perfectly known to the receiver but unknown to the transmitter. Here,

$$y = \sqrt{\rho} g x + w, \tag{2.41}$$

where ρ, x, and w have the same meaning as in Section 2.3.1, x and w are independent, and, in addition, g is a random variable that represents the fading channel gain and which is independent of x and w; see Figure 2.9(c). New independent realizations of g and w are drawn for each transmitted symbol x. The distribution of g can be arbitrary. The capacity is

$$C = \mathsf{E}\left\{\log_2\left(1 + \rho|g|^2\right)\right\}. \tag{2.42}$$

Operationally, the capacity in (2.42) only has a meaning if there is coding across all sources of randomness in the channel – including both the noise and channel gain.[1] To stress this fact, C is often called the *ergodic* capacity.

2.3.4 Fading Channel with Additive Non-Gaussian Noise and no CSI at the Receiver

We next extend the previous model to the case of an unknown channel gain and non-Gaussian noise; see Figure 2.9(d). Equation (2.41) applies; $\mathsf{E}\left\{|x|^2\right\} \leq 1$, $\mathsf{E}\{w\} = 0$, and $\mathsf{Var}\{w\} = 1$, however, w is not necessarily Gaussian. The signal x and the noise w are uncorrelated conditioned on g, $\mathsf{E}\{x^* w| \, \mathrm{g}\} = 0$, but not necessarily independent. Neither the transmitter, nor the receiver knows g. Also, g and x are independent; however, no assumption is made on the statistical relation between g and w.

To obtain a simple capacity bound, we rewrite the expression for y as follows:

$$y = \sqrt{\rho}\mathsf{E}\{g\}x + \sqrt{\rho}\left(g - \mathsf{E}\{g\}\right)x + w. \tag{2.43}$$

The receiver's lack of knowledge about g is captured by the second term, $\sqrt{\rho}\left(g - \mathsf{E}\{g\}\right)x$. A direct calculation shows that the second and third terms of (2.43) are mutually uncorrelated, and uncorrelated with x. Considering the two last terms of (2.43) as *effective noise*, the channel in (2.43) is, with appropriate normalization, equivalent to the model we treated in Section 2.3.2. Using (2.40) results in

$$C \geq \log_2\left(1 + \frac{\rho\,|\mathsf{E}\{g\}|^2}{\rho\mathsf{Var}\{g\} + 1}\right). \tag{2.44}$$

The bound in (2.44) is mainly useful when g fluctuates only slightly around its expected value $\mathsf{E}\{g\}$, so that $\mathsf{Var}\{g\}$ is small. This will be the case in many of the derivations of capacity bounds for Massive MIMO in Chapters 3 and 4. The train of reasoning can be summarized as follows: (i) Owing to the linear beamforming, each terminal sees a scalar channel with unknown gain g, and additive uncorrelated effective noise that comprises receiver noise and interference. (ii) The effect of the lack of knowledge of g, captured

[1] By contrast, in case each codeword sees only one realization of g, the capacity expression in (2.42) is irrelevant. Instead, a quantity called *outage capacity* must be considered; see, for example, [29].

in the deviation $g - \mathsf{E}\{g\}$, is treated as additional uncorrelated effective noise. By virtue of channel hardening, while g is random and unknown, it fluctuates only slightly around $\mathsf{E}\{g\}$ so this additional effective noise is small. (iii) The variances of all effective noise terms depend only on second- and fourth-order moments of Gaussian random variables, and hence can be computed in closed form.

2.3.5 Fading Channel with Non-Gaussian Noise and Side Information

The final, and most general, case of interest is that of a fading channel with non-Gaussian noise, where the receiver has access to *side information* quantified via a random variable Ω; see Figure 2.9(e). The received signal is given by (2.41) where $\mathsf{E}\left\{|x|^2\right\} \leq 1$, and w has an arbitrary distribution. The side information Ω may be correlated with g. Hence, while the receiver has no direct access to g, its knowledge of Ω may convey implicit information about g. We assume that x is independent of g and of Ω, that w has zero mean, and that x and w are uncorrelated (but not necessarily independent), conditioned on Ω, in the precise sense that

$$\begin{aligned} \mathsf{E}\{w|\Omega\} &= \mathsf{E}\{x^* w|\Omega\} \\ &= \mathsf{E}\{g^* x^* w|\Omega\} \\ &= 0. \end{aligned} \tag{2.45}$$

Then the capacity is bounded as follows:

$$C \geq \mathsf{E}\left\{\log_2\left(1 + \frac{\rho\,|\mathsf{E}\{g|\Omega\}|^2}{\rho\mathsf{Var}\{g|\Omega\} + \mathsf{Var}\{w|\Omega\}}\right)\right\}, \tag{2.46}$$

where the outer expectation is with respect to Ω.

Three special cases of (2.46) are noteworthy – in these special cases, we assume additionally that $\mathsf{Var}\{w\} = 1$:

1. In the absence of side information Ω, we revert to the case in Section 2.3.4. The bound (2.46) then reduces to (2.44), as expected.

2. If the receiver knows g so that $\Omega = g$, and w is independent of g, then (2.46) specializes to

$$C \geq \mathsf{E}\left\{\log_2\left(1 + \rho\,|g|^2\right)\right\}. \tag{2.47}$$

   The right-hand side of (2.47) is equal to (2.42). Hence, the bound (2.47) is tight in the case of Gaussian noise.

3. In the absence of fading, g is deterministic; say $g = 1$ for simplicity. Then (2.46) becomes

$$C \geq \mathsf{E}\left\{\log_2\left(1 + \frac{\rho}{\mathsf{Var}\{w|\Omega\}}\right)\right\}. \tag{2.48}$$

By applying Jensen's inequality (see Section C.1), specifically (C.3), to (2.48), we find that

$$\begin{aligned} C &\geq \log_2\left(1 + \frac{\rho}{\mathsf{Var}\{w\}}\right) \\ &= \log_2(1+\rho). \end{aligned} \tag{2.49}$$

The bound in (2.49) coincides with the bound derived in Section 2.3.2, as expected. Moreover, comparing with (2.38), we see that the bound is tight in the special case of Gaussian noise.

2.4 Summary of Key Points

The key points of this chapter are the following:

- A *coherence interval* is a time-frequency space during which the channel is substantially time-invariant and frequency-flat, so that its effect is well modeled as multiplication by a complex-valued scalar gain. The duration of a coherence interval is the channel coherence time T_c, and the bandwidth of a coherence interval is the channel coherence bandwidth B_c. Each coherence interval contains $\tau_c = B_c T_c$ complex-valued samples, and is split into an uplink and a downlink part. Table 2.1 shows some typical values of B_c, T_c, and τ_c.

  If OFDM modulation is used, two important quantities are the number of subcarriers over which the channel frequency response is approximately constant, N_{smooth}, and the number of OFDM symbols in a slot, N_{slot}. If the slot duration is equal to the channel coherence time, then each coherence interval spans N_{smooth} subcarriers and N_{slot} consecutive OFDM symbols. For this case, Figure 2.6 shows a possible mapping from OFDM symbols and subcarriers onto the samples in a coherence interval.

- In a system with L cells, K terminals per cell, and where each cell is served by an M-element base station array, propagation is modeled as follows. On the uplink, if the K terminals in the l'th cell collectively transmit the K-vector $\boldsymbol{x}_{l'}$, the base station in the lth cell observes the M-vector

$$\boldsymbol{y}_l = \sqrt{\rho_{\text{ul}}}\boldsymbol{G}_l^l \boldsymbol{x}_l + \sqrt{\rho_{\text{ul}}} \sum_{\substack{l'=1 \\ l' \neq l}}^{L} \boldsymbol{G}_{l'}^l \boldsymbol{x}_{l'} + \boldsymbol{w}_l, \tag{2.50}$$

where $\boldsymbol{w}_l$ is a vector of receiver noise and the (m, k)th element of $\boldsymbol{G}_{l'}^l$, denoted by $g_{l'k}^{lm}$, contains the channel gain between the kth terminal in the l'th cell and the mth base station antenna in the lth cell. On the downlink, if the l'th base station transmits the M-vector $\boldsymbol{x}_{l'}$ then the terminals in the lth cell receive

$$\boldsymbol{y}_l = \sqrt{\rho_{\text{dl}}}\boldsymbol{G}_l^{l\text{T}} \boldsymbol{x}_l + \sqrt{\rho_{\text{dl}}} \sum_{\substack{l'=1 \\ l' \neq l}}^{L} \boldsymbol{G}_l^{l'\text{T}} \boldsymbol{x}_{l'} + \boldsymbol{w}_l, \tag{2.51}$$

where again, $\boldsymbol{w}_l$ is receiver noise. In (2.50) and (2.51), ρ_{ul} and ρ_{dl} have the operational interpretation of SNR.

Each channel coefficient can be decomposed as

$$g_{l'k}^{lm} = \sqrt{\beta_{l'k}^l} h_{l'k}^{lm}, \tag{2.52}$$

where $\beta_{l'k}^l$ represents the attenuation due to large-scale fading (that includes path loss and shadow fading), and $h_{l'k}^{lm}$ represents the effect of the small-scale fading.

- In Massive MIMO, we show in Chapters 3 and 4 that the effective channel to each terminal is a scalar, point-to-point channel. Ergodic capacity (bounds) for the different scalar point-to-point channels illustrated in Figure 2.9 are summarized in Table 2.3.

Deterministic channel with additive Gaussian noise; see Section 2.3.1 and Figure 2.9(a)	$C = \log_2(1+\rho)$, if x and w are independent
Deterministic channel with additive non-Gaussian noise; see Section 2.3.2 and Figure 2.9(b)	$C \geq \log_2(1+\rho)$, if $\mathsf{E}\{x^*w\} = 0$
Fading channel with additive Gaussian noise and perfect CSI at the receiver; see Section 2.3.3 and Figure 2.9(c)	$C = \mathsf{E}\left\{\log_2\left(1+\rho\,\lvert g\rvert^2\right)\right\}$, if x, w and g are mutually independent
Fading channel with additive non-Gaussian noise and no CSI at the receiver; see Section 2.3.4 and Figure 2.9(d)	$C \geq \log_2\left(1+\frac{\rho\,\lvert\mathsf{E}\{g\}\rvert^2}{\rho\mathsf{Var}\{g\}+1}\right)$, if $\mathsf{E}\{x^*w\} = 0$
Fading channel with additive non-Gaussian noise and side information; see Section 2.3.5 and Figure 2.9(e)	$C \geq \mathsf{E}\left\{\log_2\left(1+\frac{\rho\,\lvert\mathsf{E}\{g\lvert\Omega\}\rvert^2}{\rho\mathsf{Var}\{g\lvert\Omega\}+\mathsf{E}\{\lvert w\rvert^2\lvert\Omega\}}\right)\right\}$, if $\mathsf{E}\{w\lvert\Omega\} = \mathsf{E}\{x^*w\lvert\Omega\} = \mathsf{E}\{g^*x^*w\lvert\Omega\} = 0$

Table 2.3. Capacity bounds for the scalar point-to-point channels. In all cases, $\mathsf{E}\{w\} = 0$ and x is independent of g and Ω, but no other assumptions (other than those explicitly stated) on statistical independence are made.

Chapter 3

SINGLE-CELL SYSTEMS

This chapter treats the single-cell scenario of Section 2.2.1, where a base station uses an array of M antennas to communicate simultaneously with K active terminals. A great deal of Massive MIMO phenomenology surfaces in this scenario: the effects of noise, channel non-orthogonality, and channel estimation errors; the details of multiplexing and de-multiplexing; near/far effects; and the significance of power control.

Throughout the chapter, we consider only zero-forcing and maximum-ratio processing. While there are somewhat better performing alternatives: MMSE on the uplink, and suitably optimized regularized zero-forcing on the downlink [30], there are no closed-form non-asymptotic expressions available for their performance. Moreover, zero-forcing and maximum-ratio themselves tend to be optimal under high- and low-SINR conditions respectively.

3.1 Uplink Pilots and Channel Estimation

Learning the channel at the base station is a critical operation. As we have seen, a wideband channel can be decomposed into coherence intervals of duration T_c seconds and bandwidth B_c Hz. Every such interval offers $\tau_c = B_c T_c$ independent uses of a frequency-flat channel as modeled in Section 2.2.1. Figure 2.3(b) illustrates the three activities that occupy each coherence interval: uplink data transmission, uplink pilot transmission, and downlink data transmission. In every coherence interval, the terminals use τ_p of the τ_c available samples to transmit pilots that are known at both ends of the link, and from which the base station estimates the channels.

3.1.1 Orthogonal Pilots

Each coherence interval must host K pilot waveforms, and in order for them not to interfere, they have to be mutually orthogonal. Henceforth, we assume that the terminals are assigned mutually orthogonal pilot sequences of length τ_{p}, where $\tau_{\mathrm{c}} \geq \tau_{\mathrm{p}} \geq K$. Any set of orthogonal pilots with the same energies yield the same performance. The significance of τ_{p} is to quantify how much energy each terminal spends on pilots in each coherence interval. In principle, any τ_{p} samples in the uplink part of the coherence interval can be used for pilots. In practice, transmitters are typically peak-power limited, so constant-modulus signals, such as orthogonal sinewaves, make ideal pilots.

We assign the kth terminal a pilot sequence denoted by a $\tau_{\mathrm{p}} \times 1$ vector $\boldsymbol{\phi}_k$, which is the kth column of a $\tau_{\mathrm{p}} \times K$ unitary matrix, such that $\tau_{\mathrm{p}} \geq K$ and

$$\boldsymbol{\Phi}^{\mathrm{H}}\boldsymbol{\Phi} = \boldsymbol{I}_K. \tag{3.1}$$

Collectively, the terminals then transmit a $K \times \tau_{\mathrm{p}}$ signal,

$$\boldsymbol{X}_{\mathrm{p}} = \sqrt{\tau_{\mathrm{p}}}\boldsymbol{\Phi}^{\mathrm{H}}, \tag{3.2}$$

which is normalized so that each terminal expends a total energy that is equal to the duration of the pilot sequence,

$$\tau_{\mathrm{p}}\boldsymbol{\phi}_k^{\mathrm{H}}\boldsymbol{\phi}_k = \tau_{\mathrm{p}}. \tag{3.3}$$

3.1.2 De-Spreading of the Received Pilot Signal

The pilot signals propagate through the uplink channel. The base station receives the $M \times \tau_{\mathrm{p}}$ signal,

$$\begin{aligned} \boldsymbol{Y}_{\mathrm{p}} &= \sqrt{\rho_{\mathrm{ul}}}\boldsymbol{G}\boldsymbol{X}_{\mathrm{p}} + \boldsymbol{W}_{\mathrm{p}} \\ &= \sqrt{\tau_{\mathrm{p}}\rho_{\mathrm{ul}}}\boldsymbol{G}\boldsymbol{\Phi}^{\mathrm{H}} + \boldsymbol{W}_{\mathrm{p}}, \end{aligned} \tag{3.4}$$

where the entries of the $M \times \tau_{\mathrm{p}}$ receiver noise matrix, $\boldsymbol{W}_{\mathrm{p}}$, are i.i.d. $\mathcal{CN}(0, 1)$.

The base station performs a de-spreading operation by correlating the received signals with each of the K pilot sequences. This is equivalent to right-multiplying the received signal matrix by the matrix of pilots, yielding

$$\begin{aligned} \boldsymbol{Y}'_{\mathrm{p}} &= \boldsymbol{Y}_{\mathrm{p}}\boldsymbol{\Phi} \\ &= \sqrt{\tau_{\mathrm{p}}\rho_{\mathrm{ul}}}\boldsymbol{G}\boldsymbol{\Phi}^{\mathrm{H}}\boldsymbol{\Phi} + \boldsymbol{W}_{\mathrm{p}}\boldsymbol{\Phi} \\ &= \sqrt{\tau_{\mathrm{p}}\rho_{\mathrm{ul}}}\boldsymbol{G} + \boldsymbol{W}'_{\mathrm{p}}, \end{aligned} \tag{3.5}$$

where $\boldsymbol{W}'_{\mathrm{p}} = \boldsymbol{W}_{\mathrm{p}}\boldsymbol{\Phi}$ is an $M \times K$ noise matrix, whose entries are again i.i.d. $\mathrm{CN}(0, 1)$ because they are related to the original Gaussian noise matrix by a unitary multiplication.

No information is lost in the de-spreading operation because multiplication of the received pilot signal by any vector in the orthogonal complement of $\boldsymbol{\Phi}$ would only result in still another noise matrix that is statistically independent of both $\boldsymbol{G}$ and $\boldsymbol{W}'_{\mathrm{p}}$.

3.1.3 MMSE Channel Estimation

After de-spreading, the base station has a noisy version (3.5) of the channel matrix. Under the assumption of independent Rayleigh fading, the elements of the channel matrix and the noise matrix are statistically independent. Hence, channel estimation decouples over both the base station antennas and the terminals, and it is sufficient to consider only the (m, k)th component of (3.5),

$$\left[\boldsymbol{Y}'_{\mathrm{p}}\right]_{mk} = \sqrt{\tau_{\mathrm{p}}\rho_{\mathrm{ul}}}\, g_k^m + \left[\boldsymbol{W}'_{\mathrm{p}}\right]_{mk}. \tag{3.6}$$

By assumption, the large-scale fading coefficients are known, so the prior distribution of g_k^m, $\mathrm{CN}(0, \beta_k)$, is also known. The MMSE estimator is

$$\begin{aligned}\hat{g}_k^m &= \mathsf{E}\left\{g_k^m | \boldsymbol{Y}_{\mathrm{p}}\right\} \\ &= \mathsf{E}\left\{g_k^m | \boldsymbol{Y}'_{\mathrm{p}}\right\} \\ &= \frac{\sqrt{\tau_{\mathrm{p}}\rho_{\mathrm{ul}}}\beta_k}{1 + \tau_{\mathrm{p}}\rho_{\mathrm{ul}}\beta_k}\left[\boldsymbol{Y}'_{\mathrm{p}}\right]_{mk}.\end{aligned} \tag{3.7}$$

The mean-square of the channel estimate is denoted by γ_k and given by

$$\begin{aligned}\gamma_k &= \mathsf{E}\left\{|\hat{g}_k^m|^2\right\} \\ &= \frac{\tau_{\mathrm{p}}\rho_{\mathrm{ul}}\beta_k^2}{1 + \tau_{\mathrm{p}}\rho_{\mathrm{ul}}\beta_k}.\end{aligned} \tag{3.8}$$

In (3.8), γ_k is the same for all m, since the channels for each of the antennas are statistically identical. The channel estimation error is denoted by

$$\tilde{g}_k^m = \hat{g}_k^m - g_k^m, \tag{3.9}$$

and the mean-square estimation error is

$$\begin{aligned}\mathsf{E}\left\{|\tilde{g}_k^m|^2\right\} &= \frac{\beta_k}{1 + \tau_{\mathrm{p}}\rho_{\mathrm{ul}}\beta_k} \\ &= \beta_k - \gamma_k.\end{aligned} \tag{3.10}$$

The channel estimation error is uncorrelated with both the channel estimate and the pilot signal,

$$\mathsf{E}\left\{\tilde{g}_k^m \left(\hat{g}_k^m\right)^*\right\} = 0, \tag{3.11}$$

$$\mathsf{E}\left\{\tilde{g}_k^m \left[\boldsymbol{Y}_{\mathrm{p}}'\right]_{mk}^*\right\} = 0. \tag{3.12}$$

The estimation error $\tilde{g}_k^m$ and the estimate $\hat{g}_k^m$ are jointly Gaussian distributed, so the fact that they are uncorrelated implies that they are also statistically independent, something that we will require later.

Note that the estimator defined by (3.7) is the same for every antenna index, as are the mean-square channel estimate (3.8) and the mean-square error (3.10). The estimate of the channel to the kth terminal is denoted by an $M \times 1$ vector $\hat{\boldsymbol{g}}_k = [\hat{g}_k^1, \ldots, \hat{g}_k^M]^{\mathrm{T}}$, which, to facilitate subsequent derivations, is expressed in normalized form as follows:

$$\hat{\boldsymbol{g}}_k = \sqrt{\gamma_k}\boldsymbol{z}_k, \tag{3.13}$$

where the components of $\boldsymbol{z}_k$ are i.i.d. CN(0, 1). Using (3.13), we write the estimate of the complete channel matrix to the K terminals, $\boldsymbol{G}$, in normalized form as

$$\hat{\boldsymbol{G}} = \boldsymbol{Z}\boldsymbol{D}_{\boldsymbol{\gamma}}^{1/2}, \tag{3.14}$$

where $\boldsymbol{\gamma} = [\gamma_1, \ldots, \gamma_K]^{\mathrm{T}}$ is a $K \times 1$ vector of mean-square channel estimates, and $\boldsymbol{Z}$ is an $M \times K$ matrix whose entries are i.i.d. CN(0, 1).

3.2 Uplink Data Transmission

All of the complexity in uplink data transfer resides in the base station. The terminals merely weight their respective symbols by power control coefficients, and then synchronously transmit the weighted symbols. In the analysis to follow, the only assumption about the statistical distribution of the symbols is that they are uncorrelated and have zero mean. However, in a practical implementation the symbols would likely come from a QAM constellation; hence, we will denote them by $\{q_k\}$. The base station receives a signal from each antenna, and it processes these signals through a linear decoding operation, either zero-forcing or maximum-ratio processing. Here *linear decoding* refers to the operation of recovering the transmitted signal, q_k. The receiver subsequently has to perform error-correction decoding.

Power control is important in Massive MIMO in order to achieve uniformly good service, and to prevent terminals having strong channels from interfering excessively with less fortunate terminals. All power control activity is *slow*; that is, the power control coefficients depend only on the large-scale fading coefficients $\{\beta_k\}$. In particular, this implies that the power control coefficients are constant with respect to frequency and that they need to be updated only at infrequent intervals. Throughout this book, we assume, mostly for simplicity, that power control is applied only to data transmission, while the pilots are always transmitted at maximum possible power.

In more detail, the kth terminal transmits a weighted symbol,

$$x_k = \sqrt{\eta_k} q_k, \tag{3.15}$$

where η_k is a power control coefficient that satisfies $0 \leq \eta_k \leq 1$. The symbols $\{q_k\}$ have zero mean and unit variance, and they are uncorrelated,

$$\mathsf{E}\left\{\boldsymbol{q}\boldsymbol{q}^{\mathrm{H}}\right\} = \boldsymbol{I}_K, \tag{3.16}$$

where $\boldsymbol{q} = [q_1, \ldots, q_K]^{\mathrm{T}}$. The signal received by the mth base station antenna is a linear combination of the signals transmitted by all terminals (see Section 2.2.1),

$$\begin{aligned} y_m &= \sqrt{\rho_{\mathrm{ul}}} \sum_{k=1}^{K} g_k^m x_k + w_m \\ &= \sqrt{\rho_{\mathrm{ul}}} \sum_{k=1}^{K} g_k^m \sqrt{\eta_k} q_k + w_m, \end{aligned} \tag{3.17}$$

where w_m is receiver noise, independent over the antennas and with distribution $\mathrm{CN}(0, 1)$. Equivalently, the complete $M \times 1$ received signal is

$$\boldsymbol{y} = \sqrt{\rho_{\mathrm{ul}}} \boldsymbol{G} \boldsymbol{D}_{\boldsymbol{\eta}}^{1/2} \boldsymbol{q} + \boldsymbol{w}, \tag{3.18}$$

where $\boldsymbol{\eta} = [\eta_1, \ldots, \eta_K]^{\mathrm{T}}$.

The base station does not know the actual channel, but it does know the channel estimate that it derived from the uplink pilots. Within (3.18), we replace $\boldsymbol{G}$ by $\hat{\boldsymbol{G}} - \tilde{\boldsymbol{G}}$, where $\tilde{\boldsymbol{G}} = \hat{\boldsymbol{G}} - \boldsymbol{G}$ is a matrix of channel estimation errors, and in turn we replace the channel estimate by its normalized version (3.14). This yields

$$\begin{aligned} \boldsymbol{y} &= \sqrt{\rho_{\mathrm{ul}}} \hat{\boldsymbol{G}} \boldsymbol{D}_{\boldsymbol{\eta}}^{1/2} \boldsymbol{q} + \left(\boldsymbol{w} - \sqrt{\rho_{\mathrm{ul}}} \tilde{\boldsymbol{G}} \boldsymbol{D}_{\boldsymbol{\eta}}^{1/2} \boldsymbol{q}\right) \\ &= \sqrt{\rho_{\mathrm{ul}}} \boldsymbol{Z} \boldsymbol{D}_{\boldsymbol{\gamma}}^{1/2} \boldsymbol{D}_{\boldsymbol{\eta}}^{1/2} \boldsymbol{q} + \left(\boldsymbol{w} - \sqrt{\rho_{\mathrm{ul}}} \tilde{\boldsymbol{G}} \boldsymbol{D}_{\boldsymbol{\eta}}^{1/2} \boldsymbol{q}\right), \end{aligned} \tag{3.19}$$

which expresses the received signal as if the transmitted signals passed through a known channel followed by the addition of the quantities in parentheses that constitute *effective noise*. The effective noise is uncorrelated with the desired signal term because the receiver noise $\boldsymbol{w}$ is independent of everything else, and the channel estimation error $\tilde{\boldsymbol{G}}$ and the channel estimate $\hat{\boldsymbol{G}}$ are uncorrelated; see (3.11). For subsequent derivations, we require the covariance of the effective noise that appears in (3.19),

$$\begin{aligned}\mathsf{Cov}\left\{\boldsymbol{w}-\sqrt{\rho_{\mathrm{ul}}}\tilde{\boldsymbol{G}}\boldsymbol{D}_{\eta}^{1/2}\boldsymbol{q}\right\} &= \boldsymbol{I}_M + \rho_{\mathrm{ul}}\mathsf{E}\left\{\tilde{\boldsymbol{G}}\boldsymbol{D}_{\eta}\tilde{\boldsymbol{G}}^{\mathrm{H}}\right\} \\ &= \boldsymbol{I}_M + \rho_{\mathrm{ul}}\sum_{k'=1}^{K}\eta_{k'}\mathsf{E}\left\{\tilde{\boldsymbol{g}}_{k'}\tilde{\boldsymbol{g}}_{k'}^{\mathrm{H}}\right\} \\ &= \left(1+\rho_{\mathrm{ul}}\sum_{k'=1}^{K}(\beta_{k'}-\gamma_{k'})\eta_{k'}\right)\boldsymbol{I}_M. \end{aligned} \tag{3.20}$$

The base station processes its received signal through multiplication of the M-vector $\boldsymbol{y}$ by a $K \times M$ decoding matrix $\boldsymbol{A}^{\mathrm{H}}$ that is a function of the channel estimate $\hat{\boldsymbol{G}}$. We consider two types of processing: zero-forcing and maximum-ratio.

3.2.1 Zero-Forcing

Zero-forcing nominally eliminates interference among the multiplexed signals. The decoding matrix is

$$\begin{aligned}\boldsymbol{A} &= \hat{\boldsymbol{G}}\left(\hat{\boldsymbol{G}}^{\mathrm{H}}\hat{\boldsymbol{G}}\right)^{-1}\boldsymbol{D}_{\gamma}^{1/2} \\ &= \mathbf{Z}\left(\mathbf{Z}^{\mathrm{H}}\mathbf{Z}\right)^{-1}, \end{aligned} \tag{3.21}$$

where again we have used the normalized channel estimate defined in (3.14). The result of processing the signal (3.19) is

$$\begin{aligned}\boldsymbol{A}^{\mathrm{H}}\boldsymbol{y} &= \sqrt{\rho_{\mathrm{ul}}}\left(\mathbf{Z}^{\mathrm{H}}\mathbf{Z}\right)^{-1}\mathbf{Z}^{\mathrm{H}}\mathbf{Z}\boldsymbol{D}_{\gamma}^{1/2}\boldsymbol{D}_{\eta}^{1/2}\boldsymbol{q} + \left(\mathbf{Z}^{\mathrm{H}}\mathbf{Z}\right)^{-1}\mathbf{Z}^{\mathrm{H}}\left(\boldsymbol{w}-\sqrt{\rho_{\mathrm{ul}}}\tilde{\boldsymbol{G}}\boldsymbol{D}_{\eta}^{1/2}\boldsymbol{q}\right) \\ &= \sqrt{\rho_{\mathrm{ul}}}\boldsymbol{D}_{\gamma}^{1/2}\boldsymbol{D}_{\eta}^{1/2}\boldsymbol{q} + \left(\mathbf{Z}^{\mathrm{H}}\mathbf{Z}\right)^{-1}\mathbf{Z}^{\mathrm{H}}\left(\boldsymbol{w}-\sqrt{\rho_{\mathrm{ul}}}\tilde{\boldsymbol{G}}\boldsymbol{D}_{\eta}^{1/2}\boldsymbol{q}\right). \end{aligned} \tag{3.22}$$

Therefore, the kth component of the processed signal,

$$\left[\boldsymbol{A}^{\mathrm{H}}\boldsymbol{y}\right]_k = \sqrt{\rho_{\mathrm{ul}}\gamma_k\eta_k}q_k + \left[\left(\mathbf{Z}^{\mathrm{H}}\mathbf{Z}\right)^{-1}\mathbf{Z}^{\mathrm{H}}\left(\boldsymbol{w}-\sqrt{\rho_{\mathrm{ul}}}\tilde{\boldsymbol{G}}\boldsymbol{D}_{\eta}^{1/2}\boldsymbol{q}\right)\right]_k, \tag{3.23}$$

is equal to the constant $\sqrt{\rho_{\mathrm{ul}}\gamma_k\eta_k}$ times the desired signal q_k, plus effective noise. Conditioned on $\mathbf{Z}$, the effective noise is uncorrelated with the desired signal. Since $\mathbf{Z}$ is known to the receiver, to obtain a capacity bound we can apply the result in Section 2.3.5, treating $\boldsymbol{Z}$ as side information. To that end, we need to evaluate the variance of the effective noise, conditioned on $\boldsymbol{Z}$. In view of (3.20) and the fact that $\tilde{\boldsymbol{G}}$ is statistically independent of the channel estimate $\hat{\boldsymbol{G}}$, and hence of $\boldsymbol{Z}$, we have that

$$\begin{aligned}
&\mathsf{Cov}\left\{\left(\boldsymbol{Z}^{\mathrm{H}}\boldsymbol{Z}\right)^{-1}\boldsymbol{Z}^{\mathrm{H}}\left(\boldsymbol{w}-\sqrt{\rho_{\mathrm{ul}}}\tilde{\boldsymbol{G}}\boldsymbol{D}_{\eta}^{1/2}\boldsymbol{q}\right)\middle|\boldsymbol{Z}\right\} \\
&=\left(1+\rho_{\mathrm{ul}}\sum_{k'=1}^{K}\left(\beta_{k'}-\gamma_{k'}\right)\eta_{k'}\right)\left(\boldsymbol{Z}^{\mathrm{H}}\boldsymbol{Z}\right)^{-1}\boldsymbol{Z}^{\mathrm{H}}\boldsymbol{Z}\left(\boldsymbol{Z}^{\mathrm{H}}\boldsymbol{Z}\right)^{-1} \\
&=\left(1+\rho_{\mathrm{ul}}\sum_{k'=1}^{K}\left(\beta_{k'}-\gamma_{k'}\right)\eta_{k'}\right)\left(\boldsymbol{Z}^{\mathrm{H}}\boldsymbol{Z}\right)^{-1}. \qquad (3.24)
\end{aligned}$$

It follows from (3.24) that the variance of the effective noise for the kth terminal, conditioned on $\boldsymbol{Z}$, is

$$\begin{aligned}
&\mathsf{Var}\left\{\left[\left(\boldsymbol{Z}^{\mathrm{H}}\boldsymbol{Z}\right)^{-1}\boldsymbol{Z}^{\mathrm{H}}\left(\boldsymbol{w}-\sqrt{\rho_{\mathrm{ul}}}\tilde{\boldsymbol{G}}\boldsymbol{D}_{\eta}^{1/2}\boldsymbol{q}\right)\right]_{k}\middle|\boldsymbol{Z}\right\} \\
&=\left(1+\rho_{\mathrm{ul}}\sum_{k'=1}^{K}\left(\beta_{k'}-\gamma_{k'}\right)\eta_{k'}\right)\left[\left(\boldsymbol{Z}^{\mathrm{H}}\boldsymbol{Z}\right)^{-1}\right]_{kk}. \qquad (3.25)
\end{aligned}$$

Using the result in Section 2.3.5, we obtain the following lower bound on the *instantaneous ergodic capacity* for the kth terminal:

$$C_{\mathrm{inst.},k}^{\mathrm{zf,ul}} \geq \mathsf{E}\left\{\log_2\left(1+\frac{\rho_{\mathrm{ul}}\gamma_k\eta_k}{\left(1+\rho_{\mathrm{ul}}\sum_{k'=1}^{K}\left(\beta_{k'}-\gamma_{k'}\right)\eta_{k'}\right)\left[\left(\boldsymbol{Z}^{\mathrm{H}}\boldsymbol{Z}\right)^{-1}\right]_{kk}}\right)\right\}. \qquad (3.26)$$

We call this capacity "instantaneous" to stress that the effective loss of samples due to the transmission of pilots in each coherence interval has not yet been considered. This cost will be taken into account when computing the net spectral efficiency in Section 3.6. In obtaining (3.26), we have implicitly assumed that there is coding over many coherence intervals that experience independent fading; hence, $C_{\mathrm{inst.},k}^{\mathrm{zf,ul}}$ is a bound on the ergodic capacity. Since all analysis in this book rests on that assumption, for brevity in what follows we will say "instantaneous capacity" instead of "instantaneous ergodic capacity".

Use and Then Forget CSI

Equation (3.26) is difficult to interpret because of the expectation outside the logarithm. In what follows, we derive an alternative, simpler lower bound which, in most cases, is fairly tight. The basic observation is that if the receiver neglects its knowledge of $\boldsymbol{Z}$ (called side information in Section 2.3.5), the model in (3.23) is equivalent to the scalar channel considered in Section 2.3.2. The interpretation of neglecting the knowledge of $\boldsymbol{Z}$ is that a first party uses the channel estimate to perform zero-forcing processing, and then passes the processed signal to a second party, but withholds the knowledge of the channel estimate. The second party performs error correction decoding, but treats the channel estimate as unknown.

To compute the bound, we use (3.24) and (B.1) in Appendix B to find the unconditional variance of the effective noise in (3.23),

$$\begin{aligned}
&\mathsf{Var}\left\{\left[\left(\boldsymbol{Z}^{\mathrm{H}}\boldsymbol{Z}\right)^{-1}\boldsymbol{Z}^{\mathrm{H}}\left(\boldsymbol{w}-\sqrt{\rho_{\mathrm{ul}}}\tilde{\boldsymbol{G}}\boldsymbol{D}_{\eta}^{1/2}\boldsymbol{q}\right)\right]_k\right\} \\
&= \left(1+\rho_{\mathrm{ul}}\sum_{k'=1}^{K}\left(\beta_{k'}-\gamma_{k'}\right)\eta_{k'}\right)\mathsf{E}\left\{\left[\left(\boldsymbol{Z}^{\mathrm{H}}\boldsymbol{Z}\right)^{-1}\right]_{kk}\right\} \\
&= \left(1+\rho_{\mathrm{ul}}\sum_{k'=1}^{K}\left(\beta_{k'}-\gamma_{k'}\right)\eta_{k'}\right)\frac{1}{M-K}.
\end{aligned} \tag{3.27}$$

Then using (2.40), we obtain

$$C_{\mathrm{inst.},k}^{\mathrm{zf,ul}} \geq \log_2\left(1+\mathrm{SINR}_k^{\mathrm{zf,ul}}\right), \tag{3.28}$$

where

$$\mathrm{SINR}_k^{\mathrm{zf,ul}} = \frac{(M-K)\rho_{\mathrm{ul}}\gamma_k\eta_k}{1+\rho_{\mathrm{ul}}\sum_{k'=1}^{K}\left(\beta_{k'}-\gamma_{k'}\right)\eta_{k'}}. \tag{3.29}$$

The quantity $\mathrm{SINR}_k^{\mathrm{zf,ul}}$ can be interpreted as an *effective SINR*, in the sense that the capacity bound is equal to the capacity of an additive Gaussian noise channel whose SNR equals the effective SINR.

Interpretation of the Effective SINR and Functional Block Diagram

We can interpret the effective SINR expression in (3.29) as follows:

- We call the quantity in the numerator the *coherent beamforming gain.* Note that the coherent beamforming gain is proportional to the mean-square channel estimate, γ_k, rather than to the inherent power of the channel, β_k.
- The first term in the denominator represents the unit-variance receiver noise.

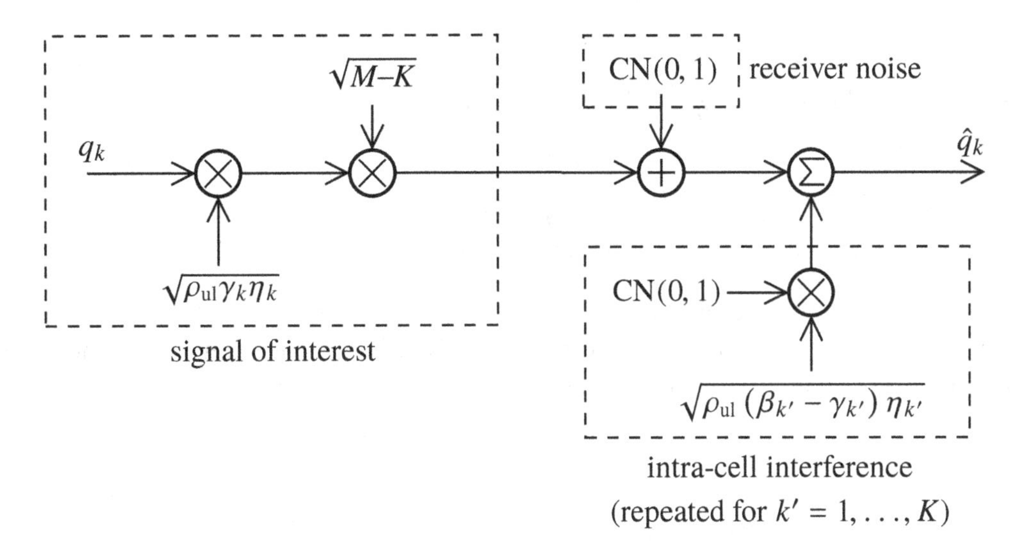

Figure 3.1. Functional block diagram for the single-cell uplink with zero-forcing processing.

- The second term in the denominator corresponds to the effects of channel estimation errors.

Throughout this chapter, for a single-cell system we use the term *intra-cell interference* to denote all terms in the denominator except for the receiver noise.

The SINR expression in (3.29) motivates an equivalent system description as a functional block diagram; see Figure 3.1. This diagram describes the passage of the original symbol transmitted by the kth terminal to its final estimate at the base station. The effective channel is scalar-valued and flat with respect to frequency.

3.2.2 Maximum-Ratio

The philosophy behind maximum-ratio processing is to amplify the signal of interest as much as possible, disregarding interference. If only one terminal were transmitting, this processing would be optimal. Here the linear decoding matrix is

$$\boldsymbol{A} = \hat{\boldsymbol{G}}\boldsymbol{D}_{\gamma}^{-1/2} = \boldsymbol{Z}. \tag{3.30}$$

The output of the maximum-ratio processing is

$$\boldsymbol{A}^{\text{H}}\boldsymbol{y} = \sqrt{\rho_{\text{ul}}}\boldsymbol{Z}^{\text{H}}\boldsymbol{Z}\boldsymbol{D}_{\gamma}^{1/2}\boldsymbol{D}_{\eta}^{1/2}\boldsymbol{q} + \boldsymbol{Z}^{\text{H}}\left(\boldsymbol{w} - \sqrt{\rho_{\text{ul}}}\tilde{\boldsymbol{G}}\boldsymbol{D}_{\eta}^{1/2}\boldsymbol{q}\right). \tag{3.31}$$

Thus,

$$\left[A^{\mathrm{H}}y\right]_k = \sqrt{\rho_{\mathrm{ul}}}z_k^{\mathrm{H}}ZD_\gamma^{1/2}D_\eta^{1/2}q + z_k^{\mathrm{H}}\left(w - \sqrt{\rho_{\mathrm{ul}}}\tilde{G}D_\eta^{1/2}q\right)$$

$$= \sqrt{\rho_{\mathrm{ul}}\gamma_k\eta_k}\|z_k\|^2 q_k + z_k^{\mathrm{H}}\left(w - \sqrt{\rho_{\mathrm{ul}}}\tilde{G}D_\eta^{1/2}q + \sum_{\substack{k'=1 \\ k'\neq k}}^{K}\sqrt{\rho_{\mathrm{ul}}\gamma_{k'}\eta_{k'}}z_{k'}q_{k'}\right). \quad (3.32)$$

There are now three sources of mutually uncorrelated effective noise: the two that figured in zero-forcing (3.23) corresponding to receiver noise and channel estimation errors, and a new term due to non-orthogonality of the estimated channel vectors. The latter has conditional covariance,

$$\mathsf{Cov}\left\{\sum_{\substack{k'=1 \\ k'\neq k}}^{K}\sqrt{\rho_{\mathrm{ul}}\gamma_{k'}\eta_{k'}}z_{k'}q_{k'}\middle| Z\right\} = \rho_{\mathrm{ul}}\sum_{\substack{k'=1 \\ k'\neq k}}^{K}\gamma_{k'}\eta_{k'}z_{k'}z_{k'}^{\mathrm{H}}. \quad (3.33)$$

Similarly to zero-forcing, we can apply the capacity bounding technique in Section 2.3.5. To do so, we first use (3.20) together with (3.33) to compute the variance of the effective noise in (3.32) conditioned on Z,

$$\mathsf{Var}\left\{z_k^{\mathrm{H}}\left(w - \sqrt{\rho_{\mathrm{ul}}}\tilde{G}D_\eta^{1/2}q + \sum_{\substack{k'=1 \\ k'\neq k}}^{K}\sqrt{\rho_{\mathrm{ul}}\gamma_{k'}\eta_{k'}}z_{k'}q_{k'}\right)\middle| Z\right\}$$

$$= \left(1 + \rho_{\mathrm{ul}}\sum_{k'=1}^{K}(\beta_{k'} - \gamma_{k'})\eta_{k'}\right)\|z_k\|^2 + \rho_{\mathrm{ul}}\sum_{\substack{k'=1 \\ k'\neq k}}^{K}\gamma_{k'}\eta_{k'}\left|z_k^{\mathrm{H}}z_{k'}\right|^2. \quad (3.34)$$

This yields the bound,

$$C_{\mathrm{inst.},k}^{\mathrm{mr,ul}} \geq \mathsf{E}\left\{\log_2\left(1 + \frac{\rho_{\mathrm{ul}}\gamma_k\eta_k\|z_k\|^2}{1 + \rho_{\mathrm{ul}}\sum_{k'=1}^{K}(\beta_{k'} - \gamma_{k'})\eta_{k'} + \rho_{\mathrm{ul}}\sum_{\substack{k'=1 \\ k'\neq k}}^{K}\gamma_{k'}\eta_{k'}\left|\frac{z_k^{\mathrm{H}}z_{k'}}{\|z_k\|}\right|^2}\right)\right\}. \quad (3.35)$$

Use and Then Forget CSI

We again obtain an inferior but more tractable bound by pretending that a first party performs maximum-ratio processing, and then passes the processed signal to a second party without CSI.

To compute the bound, we normalize the processed signal for the kth terminal in (3.32) by $1/\sqrt{M}$ and rewrite it as follows:

$$\begin{aligned}\sqrt{\frac{1}{M}}\left[A^{\mathrm{H}}y\right]_k &= \sqrt{\frac{\rho_{\mathrm{ul}}\gamma_k\eta_k}{M}}\mathsf{E}\left\{\|z_k\|^2\right\}q_k \\ &+ \sqrt{\frac{1}{M}}z_k^{\mathrm{H}}\left(w - \sqrt{\rho_{\mathrm{ul}}}\tilde{G}D_\eta^{1/2}q\right) \\ &+ \sqrt{\frac{1}{M}}z_k^{\mathrm{H}}\left(\sum_{\substack{k'=1\\k'\neq k}}^{K}\sqrt{\rho_{\mathrm{ul}}\gamma_{k'}\eta_{k'}}z_{k'}q_{k'}\right) \\ &+ \sqrt{\frac{\rho_{\mathrm{ul}}\gamma_k\eta_k}{M}}\left(\|z_k\|^2 - \mathsf{E}\left\{\|z_k\|^2\right\}\right)q_k, \end{aligned} \tag{3.36}$$

equivalent to a deterministic gain times the desired signal, plus three sources of mutually uncorrelated effective noise, which constitutes the case treated in Section 2.3.2.

We interpret the four terms in (3.36), and compute their variances, as follows:

- The first term is the desired signal. Its mean-square value is the coherent beamforming gain,

$$\frac{\rho_{\mathrm{ul}}\gamma_k\eta_k}{M}\left(\mathsf{E}\left\{\|z_k\|^2\right\}\right)^2 = M\rho_{\mathrm{ul}}\gamma_k\eta_k. \tag{3.37}$$

- The second term represents *noise and channel estimation errors*, and has variance,

$$\frac{1}{M}\mathsf{Var}\left\{z_k^{\mathrm{H}}\left(w - \sqrt{\rho_{\mathrm{ul}}}\tilde{G}D_\eta^{1/2}q\right)\right\} = 1 + \rho_{\mathrm{ul}}\sum_{k'=1}^{K}\left(\beta_{k'} - \gamma_{k'}\right)\eta_{k'}. \tag{3.38}$$

- The third term represents *channel non-orthogonality*, and has variance,

$$\begin{aligned}\frac{1}{M}\mathsf{Var}\left\{z_k^{\mathrm{H}}\sum_{\substack{k'=1\\k'\neq k}}^{K}\sqrt{\rho_{\mathrm{ul}}\gamma_{k'}\eta_{k'}}z_{k'}q_{k'}\right\} &= \frac{\rho_{\mathrm{ul}}}{M}\sum_{\substack{k'=1\\k'\neq k}}^{K}\gamma_{k'}\eta_{k'}\mathsf{E}\left\{\left|z_k^{\mathrm{H}}z_{k'}\right|^2\right\} \\ &= \rho_{\mathrm{ul}}\sum_{\substack{k'=1\\k'\neq k}}^{K}\gamma_{k'}\eta_{k'}. \end{aligned} \tag{3.39}$$

- The fourth term represents the *beamforming gain uncertainty*. This term stems from the receiver's ignorance of the effective scalar channel gain $\sqrt{\rho_{\mathrm{ul}}\gamma_k\eta_k}\|z_k\|^2$ in (3.32). Its variance follows by using the formulas in Section A.2.4:

$$\frac{\rho_{\mathrm{ul}}\gamma_k\eta_k}{M}\mathsf{Var}\left\{\left(\|z_k\|^2 - \mathsf{E}\left\{\|z_k\|^2\right\}\right) q_k\right\} = \frac{\rho_{\mathrm{ul}}\gamma_k\eta_k}{M}\left(\mathsf{E}\left\{\|z_k\|^4\right\} - \left(\mathsf{E}\left\{\|z_k\|^2\right\}\right)^2\right)$$
$$= \rho_{\mathrm{ul}}\gamma_k\eta_k. \tag{3.40}$$

The resulting capacity bound is

$$C_{\mathrm{inst.},k}^{\mathrm{mr,ul}} \geq \log_2\left(1 + \mathrm{SINR}_k^{\mathrm{mr,ul}}\right), \tag{3.41}$$

where the effective SINR is found by dividing the coherent beamforming gain in (3.37) by the sum of the variances in (3.38)–(3.40),

$$\mathrm{SINR}_k^{\mathrm{mr,ul}} = \frac{M\rho_{\mathrm{ul}}\gamma_k\eta_k}{1 + \rho_{\mathrm{ul}}\sum_{k'=1}^{K}\beta_{k'}\eta_{k'}}. \tag{3.42}$$

Functional Block Diagram

Figure 3.2 shows the functional block diagram for maximum-ratio processing. This figure should be compared with the corresponding diagram for zero-forcing processing in Figure 3.1. Maximum-ratio processing yields a coherent beamforming gain of M rather than $M - K$, since no degrees of freedom are expended on creating spatial nulls for interference suppression. However, in contrast to the zero-forcing case, the contributions to the intra-cell interference from each terminal is now proportional to the mean-square strength of the channel, β_k, rather than to the mean-square channel estimation error, $\beta_k - \gamma_k$.

3.3 Downlink Data Transmission

Downlink data transmission entails a linear precoding operation that combines the message-bearing symbols with the downlink channel estimates to create the actual signals that the array transmits.

3.3.1 Linear Precoding

Denote the $K \times 1$ vector of message-bearing symbols, as in the uplink case, by $\boldsymbol{q}$, having the covariance (3.16). The vector of transmitted signals, $\boldsymbol{x}$, is generated by first scaling the

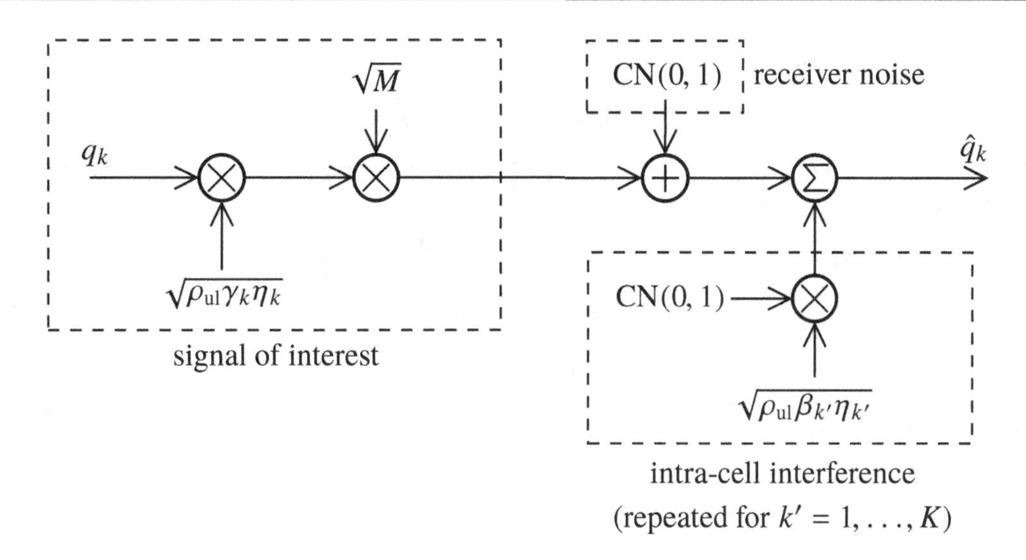

Figure 3.2. Functional block diagram for the single-cell uplink with maximum-ratio processing.

K symbols with the square root of the corresponding power control coefficients $\{\eta_k\}$, and then multiplying by an $M \times K$ precoding matrix, $\boldsymbol{A}$,

$$\begin{aligned} \boldsymbol{x} &= \boldsymbol{A}\boldsymbol{D}_{\eta}^{1/2}\boldsymbol{q} \\ &= \sum_{k=1}^{K} \sqrt{\eta_k}\boldsymbol{a}_k q_k. \end{aligned} \tag{3.43}$$

In (3.43), $\boldsymbol{a}_k$ is the kth column of $\boldsymbol{A}$. The scaling of the precoding matrix, and the choice of non-negative power control coefficients, subject to

$$\sum_{k=1}^{K} \eta_k \leq 1, \tag{3.44}$$

ensure that the total transmitted power is no greater than one: $\mathsf{E}\left\{\|\boldsymbol{x}\|^2\right\} \leq 1$. The expected power that is used for the kth terminal is

$$\mathsf{E}\left\{\|\sqrt{\eta_k}\boldsymbol{a}_k q_k\|^2\right\} = \eta_k \mathsf{E}\left\{\|\boldsymbol{a}_k\|^2\right\}. \tag{3.45}$$

Collectively, the terminals receive a $K \times 1$ signal,

$$\begin{aligned} \boldsymbol{y} &= \sqrt{\rho_{\text{dl}}}\boldsymbol{G}^{\text{T}}\boldsymbol{x} + \boldsymbol{w} \\ &= \sqrt{\rho_{\text{dl}}}\hat{\boldsymbol{G}}^{\text{T}}\boldsymbol{x} + \boldsymbol{w} - \sqrt{\rho_{\text{dl}}}\tilde{\boldsymbol{G}}^{\text{T}}\boldsymbol{x} \\ &= \sqrt{\rho_{\text{dl}}}\hat{\boldsymbol{G}}^{\text{T}}\boldsymbol{A}\boldsymbol{D}_{\eta}^{1/2}\boldsymbol{q} + \boldsymbol{w} - \sqrt{\rho_{\text{dl}}}\tilde{\boldsymbol{G}}^{\text{T}}\boldsymbol{x}. \end{aligned} \tag{3.46}$$

The signal received by the kth terminal is

$$y_k = \sqrt{\rho_{\text{dl}}}\hat{\boldsymbol{g}}_k^{\text{T}}\boldsymbol{A}\boldsymbol{D}_\eta^{1/2}\boldsymbol{q} + w_k - \sqrt{\rho_{\text{dl}}}\tilde{\boldsymbol{g}}_k^{\text{T}}\boldsymbol{x}. \tag{3.47}$$

The three terms in (3.47) are mutually uncorrelated. For subsequent use, we evaluate the variance of the sum of the last two terms,

$$\begin{aligned}\mathsf{Var}\left\{w_k - \sqrt{\rho_{\text{dl}}}\tilde{\boldsymbol{g}}_k^{\text{T}}\boldsymbol{x}\right\} &= 1 + \rho_{\text{dl}}\mathsf{E}\left\{\boldsymbol{x}^{\text{H}}\tilde{\boldsymbol{g}}_k^*\tilde{\boldsymbol{g}}_k^{\text{T}}\boldsymbol{x}\right\} \\ &= 1 + \rho_{\text{dl}}\left(\beta_k - \gamma_k\right)\mathsf{E}\left\{\|\boldsymbol{x}\|^2\right\}, \end{aligned} \tag{3.48}$$

which, significantly, depends only on the total radiated power.

As in the uplink case, we restrict ourselves to zero-forcing and maximum-ratio processing.

3.3.2 Zero-Forcing

Zero-forcing uses the precoding matrix,

$$\boldsymbol{A} = \sqrt{M-K}\boldsymbol{Z}^*\left(\boldsymbol{Z}^{\text{T}}\boldsymbol{Z}^*\right)^{-1}, \tag{3.49}$$

where $\boldsymbol{Z}$ is the normalized channel estimate defined in (3.14). We verify that the transmitted signal, (3.43), has the proper scaling,

$$\begin{aligned}\mathsf{E}\left\{\|\boldsymbol{x}\|^2\right\} &= \mathsf{E}\left\{\mathsf{Tr}\left\{\boldsymbol{A}\boldsymbol{D}_\eta\boldsymbol{A}^{\text{H}}\right\}\right\} \\ &= (M-K)\mathsf{E}\left\{\mathsf{Tr}\left\{\boldsymbol{D}_\eta\left(\boldsymbol{Z}^{\text{T}}\boldsymbol{Z}^*\right)^{-1}\right\}\right\} \\ &= (M-K)\sum_{k=1}^{K}\eta_k\mathsf{E}\left\{\left[\left(\boldsymbol{Z}^{\text{T}}\boldsymbol{Z}^*\right)^{-1}\right]_{kk}\right\} \\ &= \sum_{k=1}^{K}\eta_k, \end{aligned} \tag{3.50}$$

where in the last step we used the identity (B.1) in Appendix B. The columns of the precoding matrix, $\{\boldsymbol{a}_k\}$, are statistically identical, so (3.45) and (3.50) together imply that the power control coefficient, η_k, is in fact equal to the power expended for the kth terminal,

$$\mathsf{E}\left\{\|\sqrt{\eta_k}\boldsymbol{a}_k q_k\|^2\right\} = \eta_k. \tag{3.51}$$

The substitution of (3.49) into (3.46) gives the following received signal vector:

$$\begin{aligned}\boldsymbol{y} &= \sqrt{(M-K)\rho_{\text{dl}}}\hat{\boldsymbol{G}}^{\text{T}}\boldsymbol{Z}^*\left(\boldsymbol{Z}^{\text{T}}\boldsymbol{Z}^*\right)^{-1}\boldsymbol{D}_\eta^{1/2}\boldsymbol{q} + \boldsymbol{w} - \sqrt{\rho_{\text{dl}}}\tilde{\boldsymbol{G}}^{\text{T}}\boldsymbol{x} \\ &= \sqrt{(M-K)\rho_{\text{dl}}}\boldsymbol{D}_\gamma^{1/2}\boldsymbol{Z}^{\text{T}}\boldsymbol{Z}^*\left(\boldsymbol{Z}^{\text{T}}\boldsymbol{Z}^*\right)^{-1}\boldsymbol{D}_\eta^{1/2}\boldsymbol{q} + \boldsymbol{w} - \sqrt{\rho_{\text{dl}}}\tilde{\boldsymbol{G}}^{\text{T}}\boldsymbol{x} \\ &= \sqrt{(M-K)\rho_{\text{dl}}}\boldsymbol{D}_\gamma^{1/2}\boldsymbol{D}_\eta^{1/2}\boldsymbol{q} + \boldsymbol{w} - \sqrt{\rho_{\text{dl}}}\tilde{\boldsymbol{G}}^{\text{T}}\boldsymbol{x}. \end{aligned} \tag{3.52}$$

In particular, the kth terminal receives the signal,

$$y_k = \sqrt{(M-K)\rho_{\text{dl}}\gamma_k\eta_k}q_k + w_k - \sqrt{\rho_{\text{dl}}}\tilde{\boldsymbol{g}}_k^{\text{T}}\boldsymbol{x}. \tag{3.53}$$

Equation (3.53) expresses the received signal in terms of a deterministic gain $\sqrt{(M-K)\rho_{\text{dl}}\gamma_k\eta_k}$ multiplied by the symbol of interest, q_k, plus uncorrelated effective noise. Hence, we have the model discussed in Section 2.3.2. The variance of the effective noise follows from (3.48) and (3.50):

$$\mathsf{Var}\left\{w_k - \sqrt{\rho_{\text{dl}}}\tilde{\boldsymbol{g}}_k^{\text{T}}\boldsymbol{x}\right\} = 1 + \rho_{\text{dl}}(\beta_k - \gamma_k)\sum_{k'=1}^{K}\eta_{k'}. \tag{3.54}$$

The resulting capacity bound is

$$C_{\text{inst.},k}^{\text{zf,dl}} \geq \log_2\left(1 + \text{SINR}_k^{\text{zf,dl}}\right), \tag{3.55}$$

where the effective SINR is given by

$$\text{SINR}_k^{\text{zf,dl}} = \frac{(M-K)\rho_{\text{dl}}\gamma_k\eta_k}{1 + \rho_{\text{dl}}(\beta_k - \gamma_k)\sum_{k'=1}^{K}\eta_{k'}}. \tag{3.56}$$

Functional Block Diagram

Figure 3.3 shows the corresponding functional block diagram. This figure should be compared with the uplink zero-forcing counterpart in Figure 3.1. The numerators of the SINRs have identical form. For the downlink case, the effective noise depends only the channel estimation error for the kth terminal weighted by the total transmitted power, because the kth terminal receives power only through its own channel.

3.3.3 Maximum-Ratio

Maximum-ratio processing uses the precoding matrix,

$$\boldsymbol{A} = \frac{1}{\sqrt{M}}\boldsymbol{Z}^*. \tag{3.57}$$

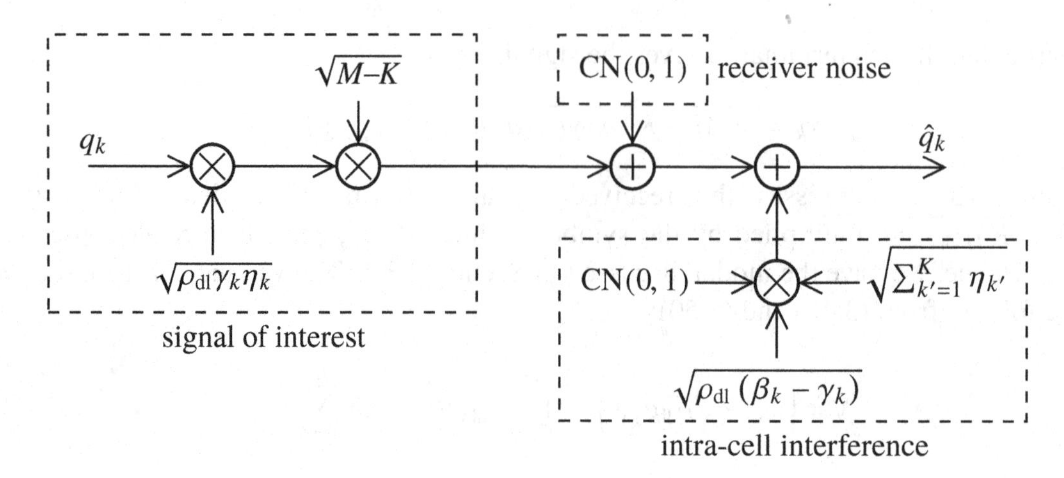

Figure 3.3. Functional block diagram for the single-cell downlink with zero-forcing processing.

We verify that the transmitted signal, (3.43), has the proper scaling,

$$\begin{aligned} \mathsf{E}\left\{\|\boldsymbol{x}\|^2\right\} &= \mathsf{E}\left\{\mathsf{Tr}\left\{\boldsymbol{A}\boldsymbol{D}_\eta\boldsymbol{A}^{\mathrm{H}}\right\}\right\} \\ &= \frac{1}{M}\mathsf{Tr}\left\{\boldsymbol{D}_\eta\mathsf{E}\left\{\boldsymbol{Z}^{\mathrm{T}}\boldsymbol{Z}^*\right\}\right\} \\ &= \frac{1}{M}\mathsf{Tr}\left\{\boldsymbol{D}_\eta M\boldsymbol{I}_K\right\} \\ &= \sum_{k=1}^{K}\eta_k. \end{aligned} \tag{3.58}$$

As in the case of zero-forcing, the expected power expended on the kth terminal is equal to η_k.

The substitution of (3.57) into (3.46) gives the received signals,

$$\begin{aligned} \boldsymbol{y} &= \sqrt{\frac{\rho_{dl}}{M}}\hat{\boldsymbol{G}}^{\mathrm{T}}\boldsymbol{Z}^*\boldsymbol{D}_\eta^{1/2}\boldsymbol{q} + \boldsymbol{w} - \sqrt{\rho_{dl}}\tilde{\boldsymbol{G}}^{\mathrm{T}}\boldsymbol{x} \\ &= \sqrt{\frac{\rho_{dl}}{M}}\boldsymbol{D}_\gamma^{1/2}\boldsymbol{Z}^{\mathrm{T}}\boldsymbol{Z}^*\boldsymbol{D}_\eta^{1/2}\boldsymbol{q} + \boldsymbol{w} - \sqrt{\rho_{dl}}\tilde{\boldsymbol{G}}^{\mathrm{T}}\boldsymbol{x} \\ &= \sqrt{\frac{\rho_{dl}}{M}}\boldsymbol{D}_\gamma^{1/2}\mathsf{E}\left\{\boldsymbol{Z}^{\mathrm{T}}\boldsymbol{Z}^*\right\}\boldsymbol{D}_\eta^{1/2}\boldsymbol{q} + \boldsymbol{w} - \sqrt{\rho_{dl}}\tilde{\boldsymbol{G}}^{\mathrm{T}}\boldsymbol{x} \\ &+ \sqrt{\frac{\rho_{dl}}{M}}\boldsymbol{D}_\gamma^{1/2}\left(\boldsymbol{Z}^{\mathrm{T}}\boldsymbol{Z}^* - \mathsf{E}\left\{\boldsymbol{Z}^{\mathrm{T}}\boldsymbol{Z}^*\right\}\right)\boldsymbol{D}_\eta^{1/2}\boldsymbol{q}. \end{aligned} \tag{3.59}$$

The kth received signal is

$$
\begin{aligned}
y_k &= \sqrt{\frac{\rho_{\rm dl}\gamma_k}{M}}\mathsf{E}\left\{\boldsymbol{z}_k^{\rm T}\boldsymbol{Z}^*\right\}\boldsymbol{D}_\eta^{1/2}\boldsymbol{q} + w_k - \sqrt{\rho_{\rm dl}}\tilde{\boldsymbol{g}}_k^{\rm T}\boldsymbol{x} + \sqrt{\frac{\rho_{\rm dl}\gamma_k}{M}}\left(\boldsymbol{z}_k^{\rm T}\boldsymbol{Z}^* - \mathsf{E}\left\{\boldsymbol{z}_k^{\rm T}\boldsymbol{Z}^*\right\}\right)\boldsymbol{D}_\eta^{1/2}\boldsymbol{q} \\
&= \sqrt{\frac{\rho_{\rm dl}\gamma_k\eta_k}{M}}\mathsf{E}\left\{\|\boldsymbol{z}_k\|^2\right\}q_k + w_k - \sqrt{\rho_{\rm dl}}\tilde{\boldsymbol{g}}_k^{\rm T}\boldsymbol{x} + \sqrt{\frac{\rho_{\rm dl}\gamma_k}{M}}\sum_{\substack{k'=1 \\ k'\neq k}}^{K}\sqrt{\eta_{k'}}\boldsymbol{z}_k^{\rm T}\boldsymbol{z}_{k'}^*q_{k'} \\
&+ \sqrt{\frac{\rho_{\rm dl}\gamma_k\eta_k}{M}}\left(\|\boldsymbol{z}_k\|^2 - \mathsf{E}\left\{\|\boldsymbol{z}_k\|^2\right\}\right)q_k. \qquad (3.60)
\end{aligned}
$$

Again we have the scalar channel model treated in Section 2.3.2. The coherent beamforming gain is equal to $M\rho_{\rm dl}\gamma_k\eta_k$, which is larger than the zero-forcing gain, as seen in (3.53). Also maximum-ratio processing results in two new sources of effective noise that are absent from zero-forcing:

- The fourth term of (3.60) represents channel non-orthogonality, similar to (3.39), with variance,

$$
\frac{\rho_{\rm dl}\gamma_k}{M}\mathsf{Var}\left\{\sum_{\substack{k'=1 \\ k'\neq k}}^{K}\sqrt{\eta_{k'}}\boldsymbol{z}_k^{\rm T}\boldsymbol{z}_{k'}^*q_{k'}\right\} = \rho_{\rm dl}\gamma_k\sum_{\substack{k'=1 \\ k'\neq k}}^{K}\eta_{k'}. \qquad (3.61)
$$

- The fifth term of (3.60) represents beamforming gain uncertainty, similar to (3.40), which has variance,

$$
\frac{\rho_{\rm dl}\gamma_k\eta_k}{M}\mathsf{Var}\left\{\left(\|\boldsymbol{z}_k\|^2 - \mathsf{E}\left\{\|\boldsymbol{z}_k\|^2\right\}\right)q_k\right\} = \rho_{\rm dl}\gamma_k\eta_k. \qquad (3.62)
$$

The sum of (3.48), (3.61), and (3.62) gives the total effective noise variance. The resulting capacity bound is

$$
C_{\text{inst.},k}^{\text{mr,dl}} \geq \log_2\left(1 + \text{SINR}_k^{\text{mr,dl}}\right), \qquad (3.63)
$$

where

$$
\text{SINR}_k^{\text{mr,dl}} = \frac{M\rho_{\rm dl}\gamma_k\eta_k}{1 + \rho_{\rm dl}\beta_k\sum_{k'=1}^{K}\eta_{k'}}. \qquad (3.64)
$$

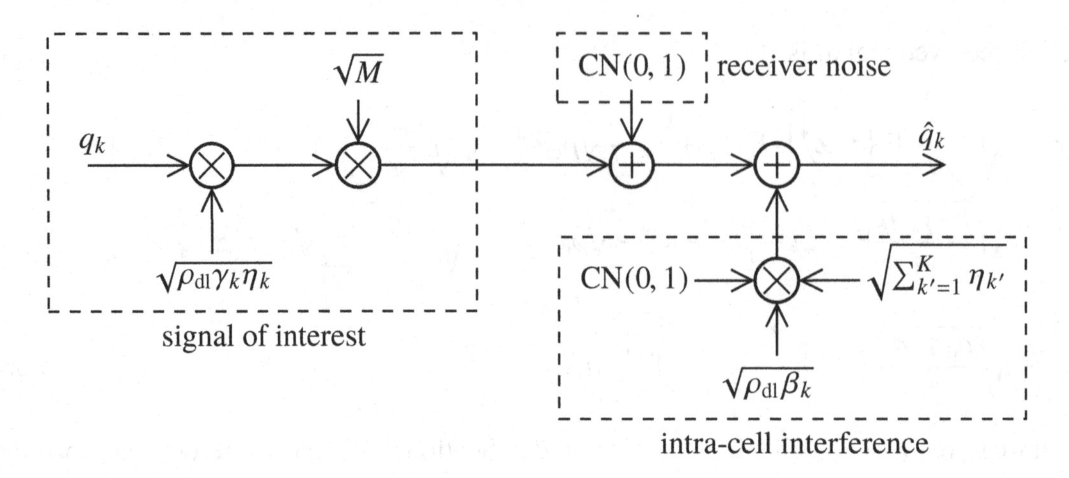

Figure 3.4. Functional block diagram for the single-cell downlink with maximum-ratio processing.

Functional Block Diagram

Figure 3.4 shows the functional block diagram, which should be compared with the block diagram for zero-forcing processing in Figure 3.3. The coherent beamforming gain for maximum-ratio transmission is proportional to M rather than $M - K$, but the effective noise is proportional to the mean-square channel rather than to the mean-square channel estimation error.

3.4 Discussion

Table 3.1 summarizes the effective SINR expressions upon which the capacity lower bounds are obtained for the four cases of uplink and downlink, zero-forcing and maximum-ratio processing. The uplink power control coefficients $\{\eta_k\}$ satisfy $0 \leq \eta_k \leq 1$. On the downlink, $\{\eta_k\}$ instead satisfy $\eta_k \geq 0$ and

$$\sum_{k=1}^{K} \eta_k \leq 1. \tag{3.65}$$

In the single-cell scenario, typically one employs full power, but in a multiple-cell scenario, at least in some cells, the total power may be less than unity.

These simple formulas are quite comprehensive, as they account for both channel estimation errors and the imperfections of the multiplexing and de-multiplexing, and they are

	Zero-Forcing	Maximum-Ratio
Uplink	$\dfrac{(M-K)\rho_{\text{ul}}\gamma_k\eta_k}{1+\rho_{\text{ul}}\sum_{k'=1}^{K}(\beta_{k'}-\gamma_{k'})\,\eta_{k'}}$	$\dfrac{M\rho_{\text{ul}}\gamma_k\eta_k}{1+\rho_{\text{ul}}\sum_{k'=1}^{K}\beta_{k'}\eta_{k'}}$
Downlink	$\dfrac{(M-K)\rho_{\text{dl}}\gamma_k\eta_k}{1+\rho_{\text{dl}}(\beta_k-\gamma_k)\sum_{k'=1}^{K}\eta_{k'}}$	$\dfrac{M\rho_{\text{dl}}\gamma_k\eta_k}{1+\rho_{\text{dl}}\beta_k\sum_{k'=1}^{K}\eta_{k'}}$

Table 3.1. Effective SINR for the kth terminal, SINR_k, in a single-cell system.

remarkably similar. It is worth noting that all effective SINR expressions were obtained without resorting to asymptotic random matrix results. Capacity lower bounds obtained from these expressions are rigorously correct, in the case of zero-forcing for any $M > K \geq 1$, and in the case of maximum-ratio processing for any $K \geq 1$ and $M \geq 1$. However, caution should be observed when using these formulas for small numbers of antennas. For example, when $M = K = 1$, the uplink maximum-ratio SINR is always less than one (omitting the redundant subscript k here),

$$\text{SINR} = \frac{\rho_{\text{ul}}\gamma\eta}{1+\rho_{\text{ul}}\beta\eta} < 1, \tag{3.66}$$

and in this regime a less tractable but stronger bound would be more useful.

Appendix D gives alternative capacity bounds for several of the scenarios treated here.

3.4.1 Interpretation of the Effective SINR Expressions

The functional block diagrams and the companion effective SINR expressions provide considerable intuitive insight:

- The frequency dependence of the channels disappears from the formulation, and only the large-scale fading coefficients appear. This is consistent with the fact that the coherent beamforming gain as well as the effects of intra-cell interference arise from the combined action of many antennas.
- The numerator represents the coherent beamforming gain for the signal designated for the kth terminal:

1. This gain is proportional to $M - K$ for zero-forcing and to M for maximum-ratio processing. The reduced gain for zero-forcing is due to the expenditure of $K - 1$ degrees of freedom for placing the signals in a null-space, and the loss of one degree of freedom due to the capacity bounding technique.
2. The factors $\rho_{\text{ul}}\gamma_k$ and $\rho_{\text{dl}}\gamma_k$ represent the effective strengths of the channel between the base station and the kth terminal, as degraded by channel estimation errors. Note that $\beta_k \geq \gamma_k$. If channel estimation were perfect, then γ_k would be equal to β_k.
3. Only the kth power control coefficient, η_k, affects the gain.

- The denominator comprises noise and intra-cell interference, whose magnitude is independent of M:

  1. The term "1" corresponds to receiver noise.
  2. The remaining term, which is convenient to think of as intra-cell interference, represents the effects of channel estimation errors, and in the case of maximum-ratio processing, channel non-orthogonality and beamforming gain uncertainty as well.
  3. On the uplink, the intra-cell interference arrives from all of the terminals and depends on the details of power control. In contrast, on the downlink the interference arrives only by the path to the kth terminal, and is proportional only to the total radiated power, i.e., the sum of the power control coefficients.
  4. For zero-forcing, the intra-cell interference depends on the mean-square channel estimation error, $\beta_k - \gamma_k$, but for maximum-ratio processing it depends on the mean-square channel, β_k. It is somewhat remarkable that for maximum-ratio processing, the combined effect of channel estimation errors, channel non-orthogonality, and beamforming gain uncertainty is independent of the quality of the channel estimates. This special result depends heavily on our assumption of independent Rayleigh fading, and it may not hold for other channel models.

- If high-quality channel estimates are available, then zero-forcing will induce considerably smaller intra-cell interference than maximum-ratio processing, yielding significantly better performance under high SNR conditions. However, as shown in Chapter 4, this advantage may evaporate in multi-cell systems because of non-coherent inter-cell interference, whose magnitude is the same for zero-forcing and maximum-ratio.

3.4.2 Implications for Power Control

The absence of frequency dependence justifies the use of power control that is independent of frequency. Hence power control for Massive MIMO is much simpler than for conventional wireless systems:

- The power control coefficients, $\{\eta_k\}$, appear linearly in both the numerator and the denominator of the SINR expressions. Hence, inequality constraints on the SINR (which are equivalent to inequality constraints on the per-terminal capacity bounds) are *linear* in the power control coefficients. As a consequence, many power control laws, for example that providing max-min throughput, are obtainable by solving linear programming problems, as shown in Chapter 5.

- Power control is a sharper tool for the uplink than for the downlink, because in the uplink the intra-cell interference depends on the detailed distribution of the power control coefficients. A terminal that has a particularly strong channel is capable of causing large interference to the other terminals. A countervailing effect is that on the downlink, power can be taken from one terminal and given to another, an option not available on the uplink.

3.4.3 Scaling Laws and Upper Bounds on the SINR

The interplay of the number of antennas, M, the number of active terminals, K, and uplink and downlink radiated powers is central to the scalability of Massive MIMO systems:

- Adding more antennas is always beneficial. The SINR is proportional to $M - K$ for zero-forcing and to M for maximum-ratio processing; however, because of its logarithmic dependence, the capacity increases at a lesser rate.

- Increased radiated power, giving proportional increases to ρ_{ul} and ρ_{dl}, is always beneficial, but under maximum-ratio processing, extra power alone has limited benefits.

  With zero-forcing, we can obtain as high performance as desired in a single-cell system by adding extra power. On the uplink, only ρ_{ul} counts in the SINR (3.29). This SINR grows without bound as ρ_{ul} grows because when $\rho_{ul} \to \infty$, (3.10) implies that $\rho_{ul}(\beta_k - \gamma_k)$ converges to a constant. Similarly, the downlink SINR (3.56) increases without bound as ρ_{dl} and ρ_{ul} simultaneously increase.

  With maximum-ratio processing, in contrast, increasing the power causes both the numerator and the denominator in the SINR expressions to increase. The quantitative implications of this become clear if we consider the arithmetic mean of the SINR (averaged over the K terminals). On the uplink, (3.42) implies that

$$\begin{aligned}\frac{1}{K}\sum_{k=1}^{K}\mathrm{SINR}_k^{\mathrm{mr,ul}} &= \frac{M}{K}\frac{\rho_{\mathrm{ul}}\sum_{k=1}^{K}\gamma_k\eta_k}{1+\rho_{\mathrm{ul}}\sum_{k'=1}^{K}\beta_{k'}\eta_{k'}} \\ &< \frac{M}{K}\frac{\sum_{k=1}^{K}\beta_k\eta_k}{\sum_{k'=1}^{K}\beta_{k'}\eta_{k'}} \\ &= \frac{M}{K}.\end{aligned} \tag{3.67}$$

Even with infinite power, the average SINR is less than the ratio of the number of antennas to the number of terminals. Similarly, on the downlink, (3.64) implies that

$$\begin{aligned}\frac{1}{K}\sum_{k=1}^{K}\mathrm{SINR}_k^{\mathrm{mr,dl}} &= \frac{M}{K}\sum_{k=1}^{K}\frac{\rho_{\mathrm{dl}}\gamma_k\eta_k}{1+\rho_{\mathrm{dl}}\beta_k\sum_{k'=1}^{K}\eta_{k'}} \\ &< \frac{M}{K}\sum_{k=1}^{K}\frac{\beta_k\eta_k}{\beta_k\sum_{k'=1}^{K}\eta_{k'}} \\ &= \frac{M}{K}.\end{aligned} \tag{3.68}$$

As a consequence, with maximum-ratio processing, the SINR saturates as the power increases without limit.

- On the downlink, doubling the number of antennas, M, permits a reduction in total radiated power by at least a factor of two with no degradation in SINR.
- On the uplink, increasing M permits a reduction in radiated power, but to a lesser extent than on the downlink, because reducing power also degrades the quality of the channel estimates. As power is reduced, eventually there is a "squaring" effect such that a doubling of M permits only a reduction in power by a factor of $\sqrt{2}$.
- Increasing the number of active terminals, K, gives a linear increase to the multiplexing gains and to the instantaneous sum capacity, but on the downlink, the available power is divided among more terminals, so the SINR decreases. For fixed

	Zero-Forcing	Maximum-Ratio
Uplink	$(M-K)\rho_{ul}\beta_k\eta_k$	$\dfrac{M\rho_{ul}\beta_k\eta_k}{1+\rho_{ul}\sum_{k'=1}^{K}\beta_{k'}\eta_{k'}}$
Downlink	$(M-K)\rho_{dl}\beta_k\eta_k$	$\dfrac{M\rho_{dl}\beta_k\eta_k}{1+\rho_{dl}\beta_k\sum_{k'=1}^{K}\eta_{k'}}$

Table 3.2. Effective SINR for the kth terminal, SINR_k, in a single-cell system, in the special case of perfect CSI at the base station ($\gamma_k = \beta_k$ for all k).

downlink power, if M is increased in proportion to K, then the same original SINR is enjoyed by more terminals. However, making K too big is eventually detrimental with respect to net spectral efficiency because of the increased time spent on training (to be discussed in more detail Section 3.6). The mobility of the terminals ultimately limits the practical scalability of Massive MIMO.

3.5 Near-Optimality of Linear Processing when $M \gg K$

Maximum-ratio and zero-forcing processing are not only scalable and computationally tractable, they also perform nearly optimally when $M \gg K$. To see this in more detail we directly compare the performance of linear processing with that of optimal processing.

In the comparison to follow, we assume that all parties know $\boldsymbol{G}$ perfectly, because exact sum capacity expressions for the uplink (multiple-access) and downlink (broadcast) channels are only available under this assumption. These sum capacities are given by (1.3) and (1.4), and derived in Section C.4. Their ergodic counterparts are obtained by taking the expectation with respect to $\boldsymbol{G}$. To find the corresponding performance with linear processing and perfect CSI, we obtain a capacity lower bound by setting $\gamma_k = \beta_k$ in the effective SINR expressions in Table 3.1, resulting in the SINRs shown in Table 3.2. We then compute $\sum_{k=1}^{K} \log_2(1 + \text{SINR}_k)$, with SINR_k taken from Table 3.2.

By comparing the capacity expressions obtained from Table 3.2 with those in (1.3) and (1.4), we make the following observations:

- For the uplink, the ergodic sum capacity is upper bounded as follows:

$$
\begin{aligned}
\mathsf{E}\left\{\log_2\left|\boldsymbol{I}_M + \rho_{\text{ul}}\boldsymbol{G}\boldsymbol{G}^{\text{H}}\right|\right\} &\overset{(a)}{=} \mathsf{E}\left\{\log_2\left|\boldsymbol{I}_K + \rho_{\text{ul}}\boldsymbol{G}^{\text{H}}\boldsymbol{G}\right|\right\} \\
&\overset{(b)}{\leq} \mathsf{E}\left\{\log_2\left(\prod_{k=1}^{K}\left[\boldsymbol{I}_K + \rho_{\text{ul}}\boldsymbol{G}^{\text{H}}\boldsymbol{G}\right]_{kk}\right)\right\} \\
&= \sum_{k=1}^{K}\mathsf{E}\left\{\log_2\left(\left[\boldsymbol{I}_K + \rho_{\text{ul}}\boldsymbol{G}^{\text{H}}\boldsymbol{G}\right]_{kk}\right)\right\} \\
&= \sum_{k=1}^{K}\mathsf{E}\left\{\log_2\left(1 + \rho_{\text{ul}}\|\boldsymbol{g}_k\|^2\right)\right\} \\
&\overset{(c)}{\leq} \sum_{k=1}^{K}\log_2\left(1 + \rho_{\text{ul}}\mathsf{E}\left\{\|\boldsymbol{g}_k\|^2\right\}\right) \\
&= \sum_{k=1}^{K}\log_2\left(1 + M\rho_{\text{ul}}\beta_k\right).
\end{aligned}
\tag{3.69}
$$

In (3.69), in (a) we used Sylvester's determinant theorem, in (b) we used the Hadamard inequality, and in (c) we used Jensen's inequality in the form (C.2). With perfect CSI and zero-forcing processing, the sum capacity is lower bounded according to

$$
\left.C_{\text{inst.,sum}}^{\text{zf,ul}}\right|_{\text{perfect CSI}} \geq \sum_{k=1}^{K}\log_2\left(1 + (M-K)\rho_{\text{ul}}\beta_k\eta_k\right).
\tag{3.70}
$$

The right-hand sides of (3.69) and (3.70) are close if all terminals use full power (so $\eta_k = 1$ for all k) and $M \gg K$. This near-optimality of zero-forcing processing is a consequence of the asymptotic orthogonality of the channel vectors in independent Rayleigh fading. See also the discussion in Chapter 7.

- Similarly, on the downlink, the ergodic sum capacity can be upper bounded, for any fixed $\{\nu_k\}$ that satisfy $\nu_k \geq 0$ and $\sum_{k=1}^{K}\nu_k \leq 1$ as follows:

$$
\mathsf{E}\left\{\log_2\left|\boldsymbol{I}_M + \rho_{\text{dl}}\boldsymbol{G}\boldsymbol{D}_{\nu}\boldsymbol{G}^{\text{H}}\right|\right\} \leq \sum_{k=1}^{K}\log_2\left(1 + M\rho_{\text{dl}}\beta_k\nu_k\right).
\tag{3.71}
$$

The sum capacity lower bound with perfect CSI and zero-forcing processing is

$$
\left.C_{\text{inst.,sum}}^{\text{zf,dl}}\right|_{\text{perfect CSI}} \geq \sum_{k=1}^{K}\log_2\left(1 + (M-K)\rho_{\text{dl}}\beta_k\eta_k\right).
\tag{3.72}
$$

Comparing the right-hand sides of (3.71) and (3.72), we see that they are close if $\nu_k = \eta_k$, and $M \gg K$.

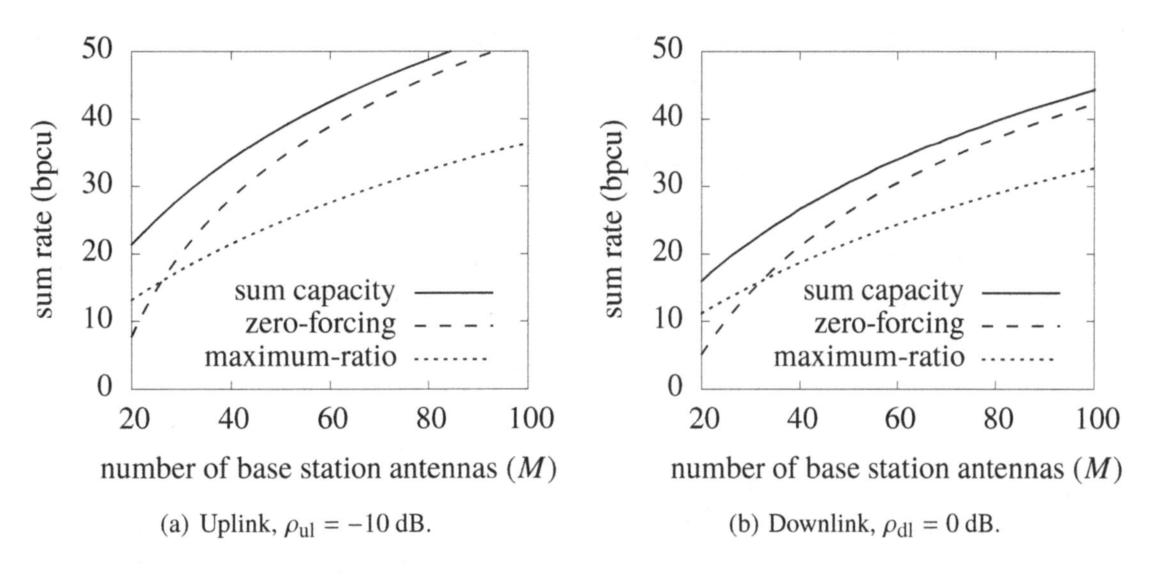

Figure 3.5. Sum capacity compared with zero-forcing and maximum-ratio as functions of M, for $K = 16$, and perfect CSI.

To give a quantitative example, suppose that the large-scale fading coefficients of all terminals are equal to unity: $\beta_k = 1$, and that no power control is applied – that is, $\eta_k = 1$ in the uplink and $\eta_k = 1/K$ in the downlink – for all k. Figure 3.5 compares the sum capacities for the uplink and downlink with the corresponding lower bounds on the sum capacity with linear processing. Results are shown for a cell with $K = 16$ terminals and SNRs of $\rho_{\text{ul}} = -10$ dB respectively $\rho_{\text{dl}} = 0$ dB. For small numbers of antennas, the $M - K$ gain factor seriously compromises zero-forcing performance. Conversely, for large M zero-forcing enjoys interference-free operation. In either case, with linear processing the sum capacity can be achieved with a small increase in numbers of antennas as compared to what is required with optimal, non-linear processing. It is worth noting that typically it would not be feasible to obtain the sum capacity shown in Figure 3.5(b) for large values of M because of the dual CSI requirement.

3.6 Net Spectral Efficiency

We obtain the (ergodic) *net spectral efficiency* for the kth terminal, $C_{\text{net},k}$, by multiplying the instantaneous (ergodic) capacity, $C_{\text{inst.},k}$, by the fraction of samples in each coherence interval that are used for transmission of payload data, that is, $1 - \tau_{\text{p}}/\tau_{\text{c}}$,

$$C_{\text{net},k} = \left(1 - \frac{\tau_{\text{p}}}{\tau_{\text{c}}}\right) C_{\text{inst.},k}. \tag{3.73}$$

The *net sum spectral efficiency* in a cell, $C_{\text{net,sum}}$, is the sum of the net spectral efficiencies of all K terminals that receive service,

$$C_{\text{net,sum}} = \left(1 - \frac{\tau_{\text{p}}}{\tau_{\text{c}}}\right) \sum_{k=1}^{K} C_{\text{inst.},k}. \tag{3.74}$$

Equations (3.73) and (3.74) apply, separately, to both the uplink and the downlink. Lower bounds on $C_{\text{inst.},k}$ are obtained from the effective SINR values in Table 3.1,

$$C_{\text{inst.},k} \geq \log_2(1 + \text{SINR}_k). \tag{3.75}$$

To obtain uplink and downlink spectral efficiencies that can be obtained simultaneously, the net spectral efficiencies in (3.73) and (3.74) must be multiplied by a factor that equals the fraction of useful samples per coherence interval that are used for uplink respectively downlink. If an equal number of useful samples are allocated for the uplink respectively the downlink, then this factor is $1/2$. To obtain *net throughputs* in b/s, all spectral efficiencies must in turn be multiplied by the total system bandwidth.

3.7 Limiting Factors: Number of Antennas and Mobility

The net sum spectral efficiency in a cell is ultimately limited either by the availability of spatial degrees of freedom – that is, the number of antennas, M, or by mobility – as quantified by the length of the channel coherence interval, τ_{c}. Both M and τ_{c} determine the number of terminals K that can be usefully multiplexed. When K becomes comparable to M, the spatial degrees of freedom are exhausted. When K increases towards τ_{c}, then the factor before the logarithm in (3.74) decreases because of the requirement that $\tau_{\text{p}} \geq K$, and eventually the whole coherence interval must be used for pilots and none of it remains for payload.

To illustrate these effects in more detail, we optimize the net sum spectral efficiency with respect to K and τ_{p}, assuming an equal split between the uplink and downlink,

$$\max_{\substack{K,\tau_{\text{p}} \\ 0 \leq K \leq \tau_{\text{p}} \leq \tau_{\text{c}}}} \frac{1}{2}\left(1 - \frac{\tau_{\text{p}}}{\tau_{\text{c}}}\right) \sum_{k=1}^{K} C_{\text{inst.},k}. \tag{3.76}$$

Figure 3.6 (for zero-forcing) and Figure 3.7 (for maximum-ratio) show the net sum spectral efficiency achieved at the optimum of (3.76) for different M and τ_{c}, visualized as contour plots. In all examples, $\beta_k = 1$ for all k, and all terminals were allocated equal power. For the sake of illustration, in this example, the optimization in (3.76) was performed independently for the uplink and the downlink. In practice, the same values of K and τ_p would have to be used in the uplink and downlink.

Figures 3.6 and 3.7 show results for the uplink and downlink for three different SNRs:

- low SNR both on uplink and downlink – $\rho_\text{ul} = \rho_\text{dl} = -5$ dB;
- low SNR on uplink and high SNR on downlink – $\rho_\text{ul} = -5$ dB and $\rho_\text{dl} = 10$ dB;
- high SNR both on uplink and on downlink – $\rho_\text{ul} = \rho_\text{dl} = 10$ dB.

Several phenomena can be observed from these figures:

- Increasing the SNRs ρ_ul and ρ_dl always helps, but beyond some point the return is diminishing owing to the logarithmic growth of capacity with power, and, for maximum-ratio processing, the persistence of interference with increasing power.
- As the uplink SNR ρ_ul increases, both the uplink and downlink capacities improve. The downlink capacities improve because the quality of the channel estimates, which are obtained on the uplink, improves. Zero-forcing benefits more from improved channel estimates than does maximum-ratio.
- When the SNR is high, zero-forcing processing yields a much higher sum capacity than maximum-ratio processing.
- Performance, both in uplink and downlink, and both for zero-forcing and maximum-ratio processing, is limited either by the number of antennas (M) or the length of the coherence interval (τ_c).

  For fixed M, the benefits of increasing τ_c eventually saturate: training becomes essentially cost-free, so one can give service to an arbitrary number of terminals. However, the effective spatial multiplexing gain is limited by M, both on the uplink and on the downlink. For zero-forcing, $K < M$ is required to obtain a non-zero capacity bound. For maximum-ratio, there is no upper limit on K, but as K grows without bound, assuming that τ_c grows correspondingly, the sum capacity approaches a finite limit. In the uplink, this saturation occurs because of persistent intra-cell interference, and in the downlink it occurs because for every doubling of K, assuming that the total power radiated by the array is fixed, the power per terminal is halved.

  Conversely, for fixed τ_c, increasing M initially increases both the coherent beamforming gain and the number of terminals that can be usefully multiplexed. However, increasing M eventually gives only a logarithmic increase in sum capacity as the coherent beamforming gain grows proportionally to M while there is no increase in multiplexing gain. The reason is that while the array offers M degrees of freedom and could spatially multiplex M terminals, there is room in each coherence interval to give at most τ_c terminals mutually orthogonal pilot sequences. Mobility becomes the limiting factor: CSI cannot be acquired for more than τ_c terminals.

While not shown in Figures 3.6 and 3.7, we have made the following additional observations, regarding the values of K and τ_p at the optimum in (3.76), for both zero-forcing and maximum-ratio processing:

- When $M \gg \tau_c$, the number of terminals that can be usefully multiplexed, K, is limited by the number of different possible orthogonal pilot sequences, which in turn is limited by the length of the coherence interval. In this regime, approximately half of the coherence interval is used for pilots; $\tau_p \approx \tau_c/2$. Also, the optimal number of terminals to multiplex is $K \approx \tau_c/2$.
- Conversely, when $\tau_c \gg M$, the number of simultaneously served terminals, K, is limited by the number of spatial degrees of freedom, M. There is room for long pilot sequences and τ_p substantially exceeds K. When the uplink SNR is high, the optimal number of terminals is $K \approx M/2$.

These observations are contingent on the assumption made in these examples, that all $\{\beta_k\}$ are equal to unity, but generalize to some extent when that assumption is relaxed. With general values of $\{\beta_k\}$, a power control policy must be selected in order for an optimization such as that in (3.76) to be meaningful. We defer further discussion of these aspects to Chapters 5 and 6.

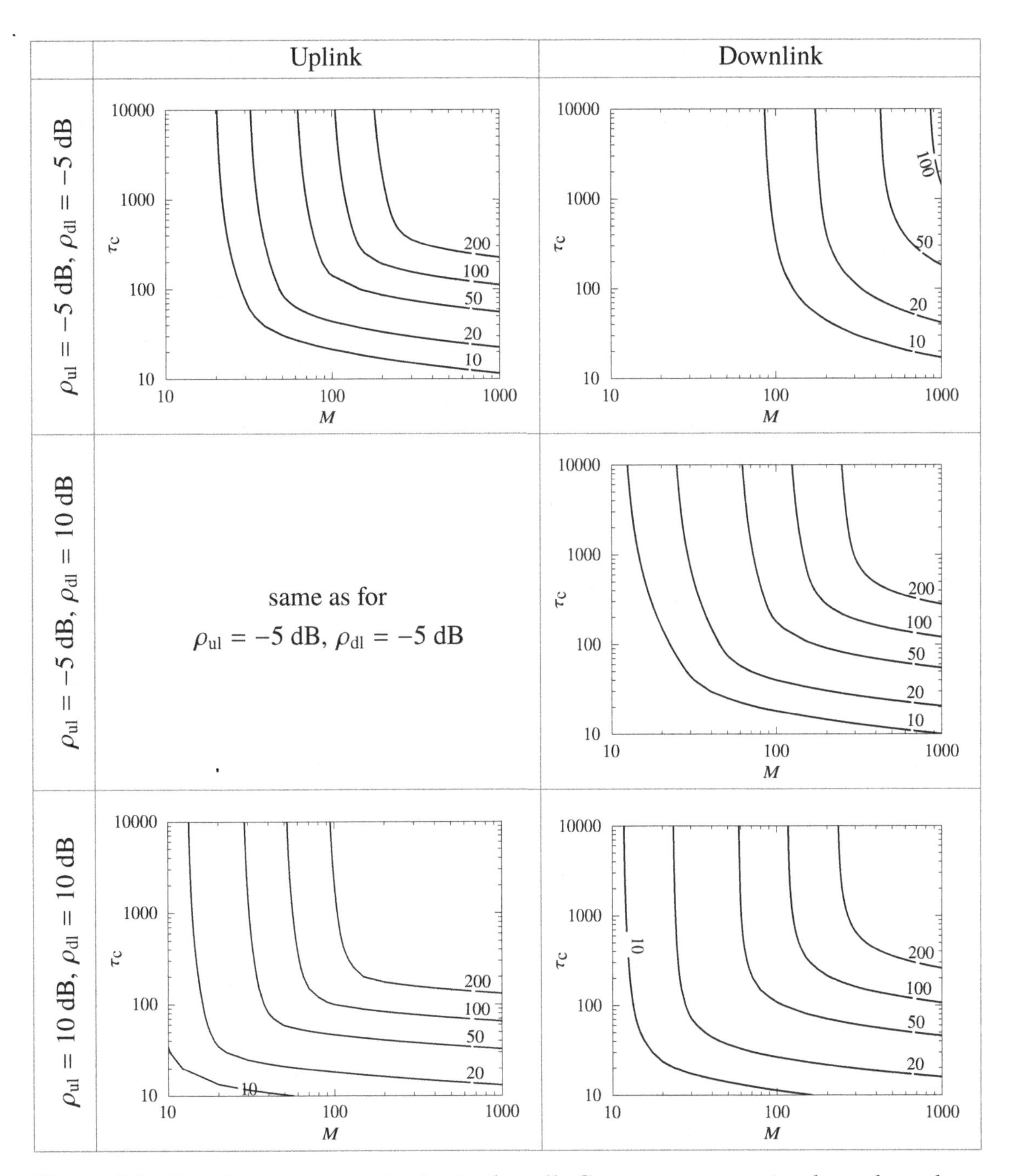

Figure 3.6. Zero-forcing processing in single-cell: Contours representing lower bounds on net sum spectral efficiency (b/s/Hz), optimized with respect to number of terminals, K, and pilot duration, τ_p; $\beta_k = 1$ for all k, and equal power expended for each terminal.

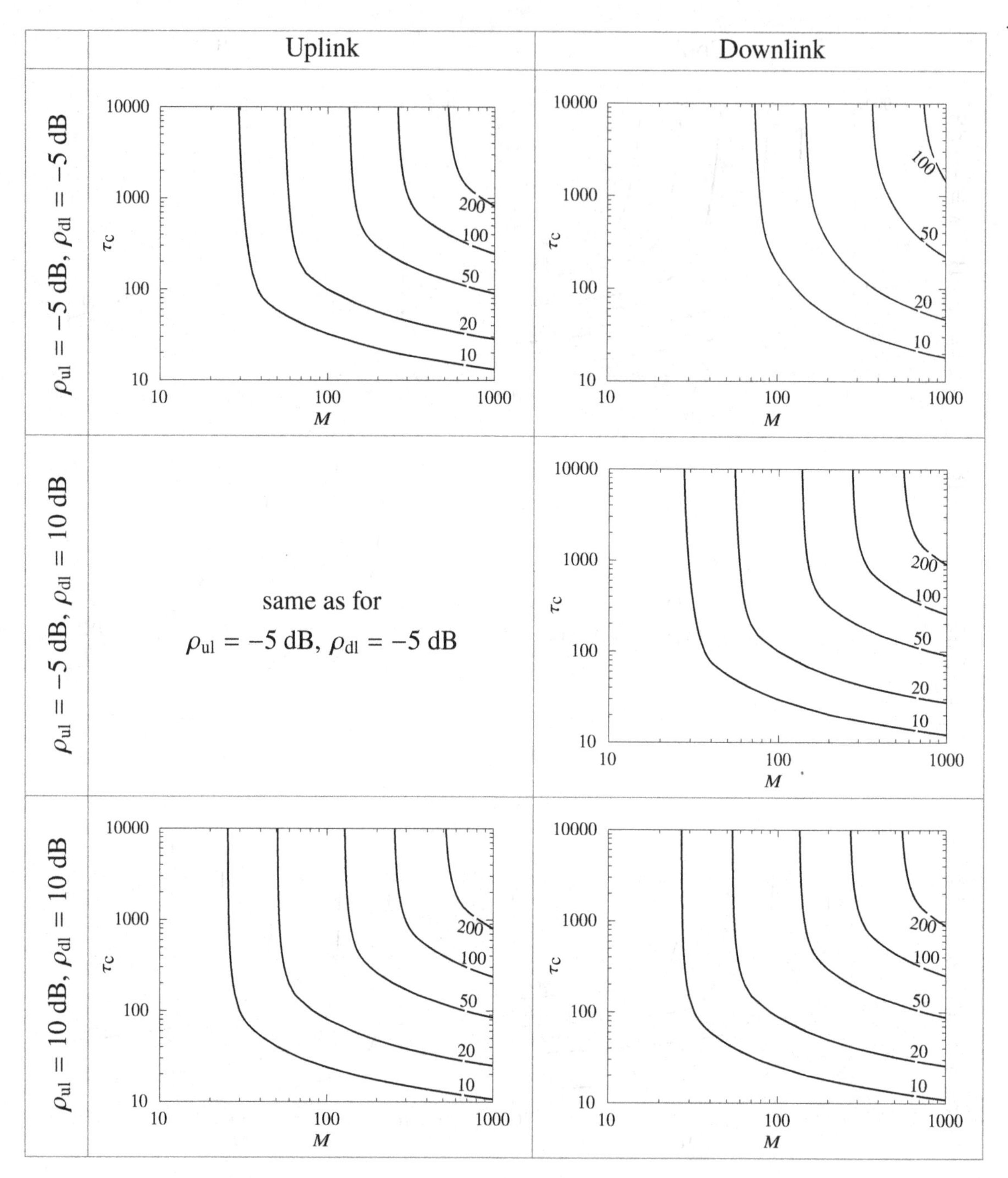

Figure 3.7. Maximum-ratio processing in single-cell: Contours representing lower bounds on net sum spectral efficiency (b/s/Hz), optimized with respect to number of terminals, K, and pilot duration, τ_p; $\beta_k = 1$ for all k, and equal power expended for each terminal.

3.8 Summary of Key Points

- Massive MIMO uses zero-forcing or maximum-ratio processing both for decoding on the uplink and precoding on the downlink. Such linear processing is nearly optimal in the typical operating regime of Massive MIMO.

- Rigorous lower bounds on the instantaneous ergodic capacity $C_{\text{inst.},k}$ for the kth terminal in a single-cell setup are given by $\log_2(1 + \text{SINR}_k)$, where SINR_k is an effective SINR, given by Table 3.1, for the uplink and the downlink and for zero-forcing and maximum-ratio processing, respectively. These bounds account for all imperfections associated with channel estimation errors, channel non-orthogonality, and beamforming gain uncertainty.

  Appendix H summarizes the notation used in Table 3.1. The power control coefficients $\{\eta_k\}$ satisfy $0 \leq \eta_k \leq 1$ for all k in the uplink. In the downlink, they satisfy the conditions $\eta_k \geq 0$ for all k, and

$$\sum_{k=1}^{K} \eta_k \leq 1. \tag{3.77}$$

  To obtain the net spectral efficiency, the instantaneous capacity has to be multiplied by the fraction of the time during which the channel is used to transmit payload data; see Section 3.6.

- The functional block diagrams in Figures 3.1–3.4 represent the effective passage of the desired signal between the terminals and the base station. In these figures, q_k is the symbol to be communicated between the array and the kth terminal, and $\hat{q}_k$ is the estimate of this symbol obtained at the receiver.

- The achievable performance in a single cell is ultimately limited either by the number of available spatial degrees of freedom – that is, the number of base station antennas, M, or by mobility as quantified via the length of the channel coherence interval, τ_c. Multiplexing to more than M terminals is not useful unless very low SINRs are desired, and learning the channel for more than τ_c terminals is impossible unless entirely new assumptions are made regarding the structure of small-scale fading.

Chapter 4

MULTI-CELL SYSTEMS

We now extend the single-cell performance analysis of Chapter 3 to the multi-cell setting as modeled in Section 2.2.2. The activities in the various cells occur synchronously, but otherwise there is no cell-to-cell cooperation, with the exception of pilot assignment and possibly power control. Each base station serves its own terminals. Data transmission activities are identical to their single-cell counterparts. The channel estimation, however, has to account for reuse of pilots in other cells.

The culmination of this chapter is Table 4.1 that summarizes effective SINRs analogous to Table 3.1 in the single-cell case. Compared with the single-cell SINRs, the multi-cell SINRs contain additional effective noise terms that correspond to inter-cell interference of two types: non-coherent interference that is independent of the number of base station antennas, and coherent interference that scales with the number of base station antennas.

Throughout this chapter, we denote the home cell by the index l. Although the terminals within each cell have mutually orthogonal pilots, some reuse of pilots from cell to cell is permitted. The assumption is that for any two distinct cells, the pilot sequences are either perfectly orthogonal from cell to cell, or are perfectly replicated. Cells that use the same pilots as the home cell cause *pilot contamination*. We call these cells *contaminating cells* and denote the set of their indices by $\mathcal{P}_l$, where by definition $\mathcal{P}_l$ also includes the home cell l. For all $l' \in \mathcal{P}_l$, the kth terminal in the l'th cell is assigned the same kth pilot sequence.

We use the capacity bounding techniques of Chapter 3, but in our treatments of the uplink we immediately adopt the "use and forget" technique, thereby skipping the forms of capacity bounds that require taking expectations of logarithms.

4.1 Uplink Pilots and Channel Estimation

During the training phase, the terminals in each cell transmit pilot sequences, as in (3.2). Upon reception of the pilots, the base station performs a de-spreading operation. Because of pilot reuse, the resulting signal is a linear combination of channel matrices from all cells that share the same pilot sequences. Within the lth cell, the resulting signal (the counterpart to (3.5)) is

$$\boldsymbol{Y}'_{\mathrm{p}l} = \sqrt{\tau_{\mathrm{p}}\rho_{\mathrm{ul}}} \sum_{l' \in \mathcal{P}_l} \boldsymbol{G}^l_{l'} + \boldsymbol{W}'_{\mathrm{p}l}, \tag{4.1}$$

where $\boldsymbol{W}'_{\mathrm{p}l}$ represents noise whose elements are i.i.d. CN(0, 1). In terms of components,

$$\left[\boldsymbol{Y}'_{\mathrm{p}l}\right]_{mk} = \sqrt{\tau_{\mathrm{p}}\rho_{\mathrm{ul}}} \sum_{l' \in \mathcal{P}_l} g^{lm}_{l'k} + \left[\boldsymbol{W}'_{\mathrm{p}l}\right]_{mk}. \tag{4.2}$$

To implement precoding and decoding, the home cell requires only an estimate of its own channel matrix, $\boldsymbol{G}^l_l$. Performance calculations and power control algorithms depend on the mean-square channel estimates to all cells.

The MMSE estimate of $g^{lm}_{l'k}$ is

$$\hat{g}^{lm}_{l'k} = \frac{\sqrt{\tau_{\mathrm{p}}\rho_{\mathrm{ul}}}\beta^l_{l'k}}{1 + \tau_{\mathrm{p}}\rho_{\mathrm{ul}} \sum_{l'' \in \mathcal{P}_l} \beta^l_{l''k}} \left[\boldsymbol{Y}'_{\mathrm{p}l}\right]_{mk}, \qquad l' \in \mathcal{P}_l. \tag{4.3}$$

Note that estimates obtained by the home cell for different values of l', but with the same terminal index, k, are perfectly correlated, which is the essence of pilot contamination. The mean-square channel estimate is denoted by $\gamma^l_{l'k}$ and given by

$$\begin{aligned} \gamma^l_{l'k} &= \mathsf{E}\left\{|\hat{g}^{lm}_{l'k}|^2\right\} \\ &= \frac{\tau_{\mathrm{p}}\rho_{\mathrm{ul}}\left(\beta^l_{l'k}\right)^2}{1 + \tau_{\mathrm{p}}\rho_{\mathrm{ul}} \sum_{l'' \in \mathcal{P}_l} \beta^l_{l''k}}, \qquad l' \in \mathcal{P}_l. \end{aligned} \tag{4.4}$$

From (4.4), it is clear that $\gamma^l_{l'k} \leq \beta^l_{l'k}$. As a consequence of pilot contamination, multi-cell channel estimates may be considerably noisier than their single-cell counterpart, as seen by comparing (4.4) with (3.8).

Let

$$\tilde{g}^{lm}_{l'k} = \hat{g}^{lm}_{l'k} - g^{lm}_{l'k} \tag{4.5}$$

be the channel estimation error. The mean-square estimation error is

$$\mathsf{E}\left\{|\tilde{g}_{l'k}^{lm}|^2\right\} = \beta_{l'k}^l - \gamma_{l'k}^l, \qquad l' \in \mathcal{P}_l, \tag{4.6}$$

independently of m. In matrix form, we represent the estimate (4.3) as follows:

$$\hat{\boldsymbol{G}}_{l'}^l = \boldsymbol{Z}^l \boldsymbol{D}_{\boldsymbol{\gamma}_{l'}^l}^{1/2}, \qquad l' \in \mathcal{P}_l, \tag{4.7}$$

where $\hat{\boldsymbol{G}}_{l'}^l$ is a matrix of channel estimates, whose (m,k)th element equals $\hat{g}_{l'k}^{lm}$, $\boldsymbol{\gamma}_{l'}^l = [\gamma_{l'1}^l, \ldots, \gamma_{l'K}^l]^{\mathrm{T}}$, and the elements of $\boldsymbol{Z}^l$ are i.i.d. CN(0, 1). In what follows, it is extremely important that $\boldsymbol{Z}^l$ has no dependence on l'.

4.2 Uplink Data Transmission

During the uplink data phase, the kth terminal in the l'th cell transmits $\sqrt{\eta_{l'k}}q_{l'k}$, where the symbols $\{q_{l'k}\}$ are mutually uncorrelated and have zero mean and unit power, and $\{\eta_{l'k}\}$ are power control coefficients that satisfy $0 \leq \eta_{l'k} \leq 1$ for all l' and k. The base station in the home cell receives the following vector-valued signal:

$$\begin{aligned}
\boldsymbol{y}_l &= \sqrt{\rho_{\mathrm{ul}}} \sum_{l' \in \mathcal{P}_l} \boldsymbol{G}_{l'}^l \boldsymbol{D}_{\boldsymbol{\eta}_{l'}}^{1/2} \boldsymbol{q}_{l'} + \sqrt{\rho_{\mathrm{ul}}} \sum_{l' \notin \mathcal{P}_l} \boldsymbol{G}_{l'}^l \boldsymbol{D}_{\boldsymbol{\eta}_{l'}}^{1/2} \boldsymbol{q}_{l'} + \boldsymbol{w}_l \\
&= \sqrt{\rho_{\mathrm{ul}}} \sum_{l' \in \mathcal{P}_l} \hat{\boldsymbol{G}}_{l'}^l \boldsymbol{D}_{\boldsymbol{\eta}_{l'}}^{1/2} \boldsymbol{q}_{l'} - \sqrt{\rho_{\mathrm{ul}}} \sum_{l' \in \mathcal{P}_l} \tilde{\boldsymbol{G}}_{l'}^l \boldsymbol{D}_{\boldsymbol{\eta}_{l'}}^{1/2} \boldsymbol{q}_{l'} + \sqrt{\rho_{\mathrm{ul}}} \sum_{l' \notin \mathcal{P}_l} \boldsymbol{G}_{l'}^l \boldsymbol{D}_{\boldsymbol{\eta}_{l'}}^{1/2} \boldsymbol{q}_{l'} + \boldsymbol{w}_l \\
&= \sqrt{\rho_{\mathrm{ul}}} \boldsymbol{Z}^l \sum_{l' \in \mathcal{P}_l} \boldsymbol{D}_{\boldsymbol{\gamma}_{l'}^l}^{1/2} \boldsymbol{D}_{\boldsymbol{\eta}_{l'}}^{1/2} \boldsymbol{q}_{l'} - \sqrt{\rho_{\mathrm{ul}}} \sum_{l' \in \mathcal{P}_l} \tilde{\boldsymbol{G}}_{l'}^l \boldsymbol{D}_{\boldsymbol{\eta}_{l'}}^{1/2} \boldsymbol{q}_{l'} + \sqrt{\rho_{\mathrm{ul}}} \sum_{l' \notin \mathcal{P}_l} \boldsymbol{G}_{l'}^l \boldsymbol{D}_{\boldsymbol{\eta}_{l'}}^{1/2} \boldsymbol{q}_{l'} + \boldsymbol{w}_l,
\end{aligned} \tag{4.8}$$

where $\tilde{\boldsymbol{G}}_{l'}^l = \hat{\boldsymbol{G}}_{l'}^l - \boldsymbol{G}_{l'}^l$ is a matrix of channel estimation errors, $\boldsymbol{\eta}_{l'} = [\eta_{l'1}, \ldots, \eta_{l'K}]^{\mathrm{T}}$, and where in the last step we utilized (4.7). We note that the four terms in (4.8) are mutually uncorrelated. Moreover, the last three terms are statistically independent of $\{\hat{\boldsymbol{G}}_{l'}^l\}$, and hence of $\boldsymbol{Z}^l$, for $l' \in \mathcal{P}_l$. Their covariance is

$$\begin{aligned}
&\mathsf{Cov}\left\{-\sqrt{\rho_{\mathrm{ul}}} \sum_{l' \in \mathcal{P}_l} \tilde{\boldsymbol{G}}_{l'}^l \boldsymbol{D}_{\boldsymbol{\eta}_{l'}}^{1/2} \boldsymbol{q}_{l'} + \sqrt{\rho_{\mathrm{ul}}} \sum_{l' \notin \mathcal{P}_l} \boldsymbol{G}_{l'}^l \boldsymbol{D}_{\boldsymbol{\eta}_{l'}}^{1/2} \boldsymbol{q}_{l'} + \boldsymbol{w}_l\right\} \\
&\qquad = \left(\rho_{\mathrm{ul}} \sum_{l' \in \mathcal{P}_l} \sum_{k'=1}^{K} \left(\beta_{l'k'}^l - \gamma_{l'k'}^l\right) \eta_{l'k'} + \rho_{\mathrm{ul}} \sum_{l' \notin \mathcal{P}_l} \sum_{k'=1}^{K} \beta_{l'k'}^l \eta_{l'k'} + 1\right) \boldsymbol{I}_M.
\end{aligned} \tag{4.9}$$

Within (4.9), note that we take an unconditional expectation with respect to $\{\boldsymbol{G}_{l'}^l\}$, despite the fact that the base station possesses estimates of these quantities in keeping with the "use and forget CSI" technique.

4.2.1 Zero-Forcing

Under zero-forcing, the home cell multiplies its received signal (4.8) by the scaled pseudo-inverse of the channel estimate. The kth component of the processed signal is as follows:

$$
\begin{aligned}
&\left[\boldsymbol{D}_{\gamma_l^l}^{1/2}\left(\hat{\boldsymbol{G}}_l^{l\mathrm{H}}\hat{\boldsymbol{G}}_l^l\right)^{-1}\hat{\boldsymbol{G}}_l^{l\mathrm{H}}\boldsymbol{y}_l\right]_k \\
&\quad=\left[\left(\mathbf{Z}^{l\mathrm{H}}\mathbf{Z}^l\right)^{-1}\mathbf{Z}^{l\mathrm{H}}\boldsymbol{y}_l\right]_k \\
&\quad=\left[\sqrt{\rho_{\mathrm{ul}}}\left(\mathbf{Z}^{l\mathrm{H}}\mathbf{Z}^l\right)^{-1}\mathbf{Z}^{l\mathrm{H}}\mathbf{Z}^l\sum_{l'\in\mathcal{P}_l}\mathbf{D}_{\gamma_{l'}^l}^{1/2}\mathbf{D}_{\eta_{l'}}^{1/2}\boldsymbol{q}_{l'}\right]_k \\
&\qquad+\left[\left(\mathbf{Z}^{l\mathrm{H}}\mathbf{Z}^l\right)^{-1}\mathbf{Z}^{l\mathrm{H}}\left(-\sqrt{\rho_{\mathrm{ul}}}\sum_{l'\in\mathcal{P}_l}\tilde{\mathbf{G}}_{l'}^l\mathbf{D}_{\eta_{l'}}^{1/2}\boldsymbol{q}_{l'}+\sqrt{\rho_{\mathrm{ul}}}\sum_{l'\notin\mathcal{P}_l}\mathbf{G}_{l'}^l\mathbf{D}_{\eta_{l'}}^{1/2}\boldsymbol{q}_{l'}+\boldsymbol{w}_l\right)\right]_k \\
&\quad=\sqrt{\rho_{\mathrm{ul}}}\sum_{l'\in\mathcal{P}_l}\sqrt{\gamma_{l'k}^l\eta_{l'k}}q_{l'k} \\
&\qquad+\left[\left(\mathbf{Z}^{l\mathrm{H}}\mathbf{Z}^l\right)^{-1}\mathbf{Z}^{l\mathrm{H}}\left(-\sqrt{\rho_{\mathrm{ul}}}\sum_{l'\in\mathcal{P}_l}\tilde{\mathbf{G}}_{l'}^l\mathbf{D}_{\eta_{l'}}^{1/2}\boldsymbol{q}_{l'}+\sqrt{\rho_{\mathrm{ul}}}\sum_{l'\notin\mathcal{P}_l}\mathbf{G}_{l'}^l\mathbf{D}_{\eta_{l'}}^{1/2}\boldsymbol{q}_{l'}+\boldsymbol{w}_l\right)\right]_k, \\
&\quad=\sqrt{\rho_{\mathrm{ul}}\gamma_{lk}^l\eta_{lk}}q_{lk}+\sqrt{\rho_{\mathrm{ul}}}\sum_{l'\in\mathcal{P}_l\setminus\{l\}}\sqrt{\gamma_{l'k}^l\eta_{l'k}}q_{l'k} \\
&\qquad+\left[\left(\mathbf{Z}^{l\mathrm{H}}\mathbf{Z}^l\right)^{-1}\mathbf{Z}^{l\mathrm{H}}\left(-\sqrt{\rho_{\mathrm{ul}}}\sum_{l'\in\mathcal{P}_l}\tilde{\mathbf{G}}_{l'}^l\mathbf{D}_{\eta_{l'}}^{1/2}\boldsymbol{q}_{l'}+\sqrt{\rho_{\mathrm{ul}}}\sum_{l'\notin\mathcal{P}_l}\mathbf{G}_{l'}^l\mathbf{D}_{\eta_{l'}}^{1/2}\boldsymbol{q}_{l'}+\boldsymbol{w}_l\right)\right]_k,
\end{aligned}
\tag{4.10}
$$

where, in the first step, we have used (4.7). The kth processed signal in (4.10) is equal to a deterministic gain multiplied by the symbols from the home-cell terminal, plus uncorrelated effective non-Gaussian noise terms. Consequently, we have an effective scalar channel of the form treated in Section 2.3.2, and can use the capacity bound derived therein.

We use (4.9) and (B.1) to evaluate the variance of the last effective noise terms in (4.10),

$$\begin{aligned}
&\mathsf{Var}\left\{\left[\left(\boldsymbol{Z}^{l\mathrm{H}}\boldsymbol{Z}^{l}\right)^{-1}\boldsymbol{Z}^{l\mathrm{H}}\left(-\sqrt{\rho_{\mathrm{ul}}}\sum_{l'\in\mathcal{P}_l}\tilde{\boldsymbol{G}}^{l}_{l'}\boldsymbol{D}^{1/2}_{\boldsymbol{\eta}_{l'}}\boldsymbol{q}_{l'}+\sqrt{\rho_{\mathrm{ul}}}\sum_{l'\notin\mathcal{P}_l}\boldsymbol{G}^{l}_{l'}\boldsymbol{D}^{1/2}_{\boldsymbol{\eta}_{l'}}\boldsymbol{q}_{l'}+\boldsymbol{w}_l\right)\right]_k\right\}\\
&=\left(\rho_{\mathrm{ul}}\sum_{l'\in\mathcal{P}_l}\sum_{k'=1}^{K}\left(\beta^{l}_{l'k'}-\gamma^{l}_{l'k'}\right)\eta_{l'k'}+\rho_{\mathrm{ul}}\sum_{l'\notin\mathcal{P}_l}\sum_{k'=1}^{K}\beta^{l}_{l'k'}\eta_{l'k'}+1\right)\mathsf{E}\left\{\left[\left(\boldsymbol{Z}^{l\mathrm{H}}\boldsymbol{Z}^{l}\right)^{-1}\right]_{kk}\right\}\\
&=\frac{1}{M-K}\left(\rho_{\mathrm{ul}}\sum_{l'\in\mathcal{P}_l}\sum_{k'=1}^{K}\left(\beta^{l}_{l'k'}-\gamma^{l}_{l'k'}\right)\eta_{l'k'}+\rho_{\mathrm{ul}}\sum_{l'\notin\mathcal{P}_l}\sum_{k'=1}^{K}\beta^{l}_{l'k'}\eta_{l'k'}+1\right),
\end{aligned} \tag{4.11}$$

where we exploited the independence between $\boldsymbol{Z}^l$ and the last three terms of (4.8). From (4.10) and (4.11), we obtain the effective SINR for the kth terminal in the lth cell, as follows:

$$\mathrm{SINR}^{\mathrm{zf,ul}}_{lk}=\frac{(M-K)\rho_{\mathrm{ul}}\gamma^{l}_{lk}\eta_{lk}}{1+\rho_{\mathrm{ul}}\sum\limits_{l'\in\mathcal{P}_l}\sum\limits_{k'=1}^{K}\left(\beta^{l}_{l'k'}-\gamma^{l}_{l'k'}\right)\eta_{l'k'}+\rho_{\mathrm{ul}}\sum\limits_{l'\notin\mathcal{P}_l}\sum\limits_{k'=1}^{K}\beta^{l}_{l'k'}\eta_{l'k'}+(M-K)\rho_{\mathrm{ul}}\sum\limits_{l'\in\mathcal{P}_l\setminus\{l\}}\gamma^{l}_{l'k}\eta_{l'k}}, \tag{4.12}$$

and the resulting capacity bound

$$C^{\mathrm{zf,ul}}_{\mathrm{inst.},lk}\geq\log_2\left(1+\mathrm{SINR}^{\mathrm{zf,ul}}_{lk}\right). \tag{4.13}$$

Interpretation of (4.12) and Functional Block Diagram

We interpret the effective SINR expression (4.12) as follows:

- The numerator represents the coherent beamforming gain of the desired signal received at the base station in the home cell.
- The first term in the denominator is the receiver noise variance.
- The second term in the denominator is due to interference from contaminating cells. This interference includes both inter-cell interference and intra-cell interference within the home cell. The terms that make up this second term scale proportionally with the mean-square channel estimation errors, but not with M.
- The third term in the denominator represents inter-cell interference from non-contaminating cells that scales proportionally with the large-scale fading coefficients, but not with M.

  We use the term *non-coherent interference* to denote the second and the third terms in the denominator.

- The fourth term in the denominator represents interference from contaminating cells, excluding the home cell itself, proportional to the mean-square channel estimates and to $M - K$. Due to the proportional scaling with $M - K$, we call this term *coherent interference*.

The functional block diagram in Figure 4.1 provides further intuition.

4.2.2 Maximum-Ratio

Under maximum-ratio processing, the home cell multiplies the received signal (4.8) by the complex conjugate of the channel estimate, resulting in

$$\begin{aligned}\left[\frac{1}{\sqrt{M}}\boldsymbol{D}_{\gamma_l^l}^{-1/2}\hat{\boldsymbol{G}}_l^{l\mathrm{H}}\boldsymbol{y}_l\right]_k &= \left[\frac{1}{\sqrt{M}}\boldsymbol{Z}^{l\mathrm{H}}\boldsymbol{y}_l\right]_k \\ &= \left[\frac{1}{\sqrt{M}}\boldsymbol{Z}^{l\mathrm{H}}\left(\sqrt{\rho_{\mathrm{ul}}}\boldsymbol{Z}^l\sum_{l'\in\mathcal{P}_l}\boldsymbol{D}_{\gamma_{l'}^l}^{1/2}\boldsymbol{D}_{\eta_{l'}}^{1/2}\boldsymbol{q}_{l'}\right.\right. \\ &\quad \left.\left. -\sqrt{\rho_{\mathrm{ul}}}\sum_{l'\in\mathcal{P}_l}\tilde{\boldsymbol{G}}_{l'}^l\boldsymbol{D}_{\eta_{l'}}^{1/2}\boldsymbol{q}_{l'} + \sqrt{\rho_{\mathrm{ul}}}\sum_{l'\notin\mathcal{P}_l}\boldsymbol{G}_{l'}^l\boldsymbol{D}_{\eta_{l'}}^{1/2}\boldsymbol{q}_{l'} + \boldsymbol{w}_l\right)\right]_k. \end{aligned} \tag{4.14}$$

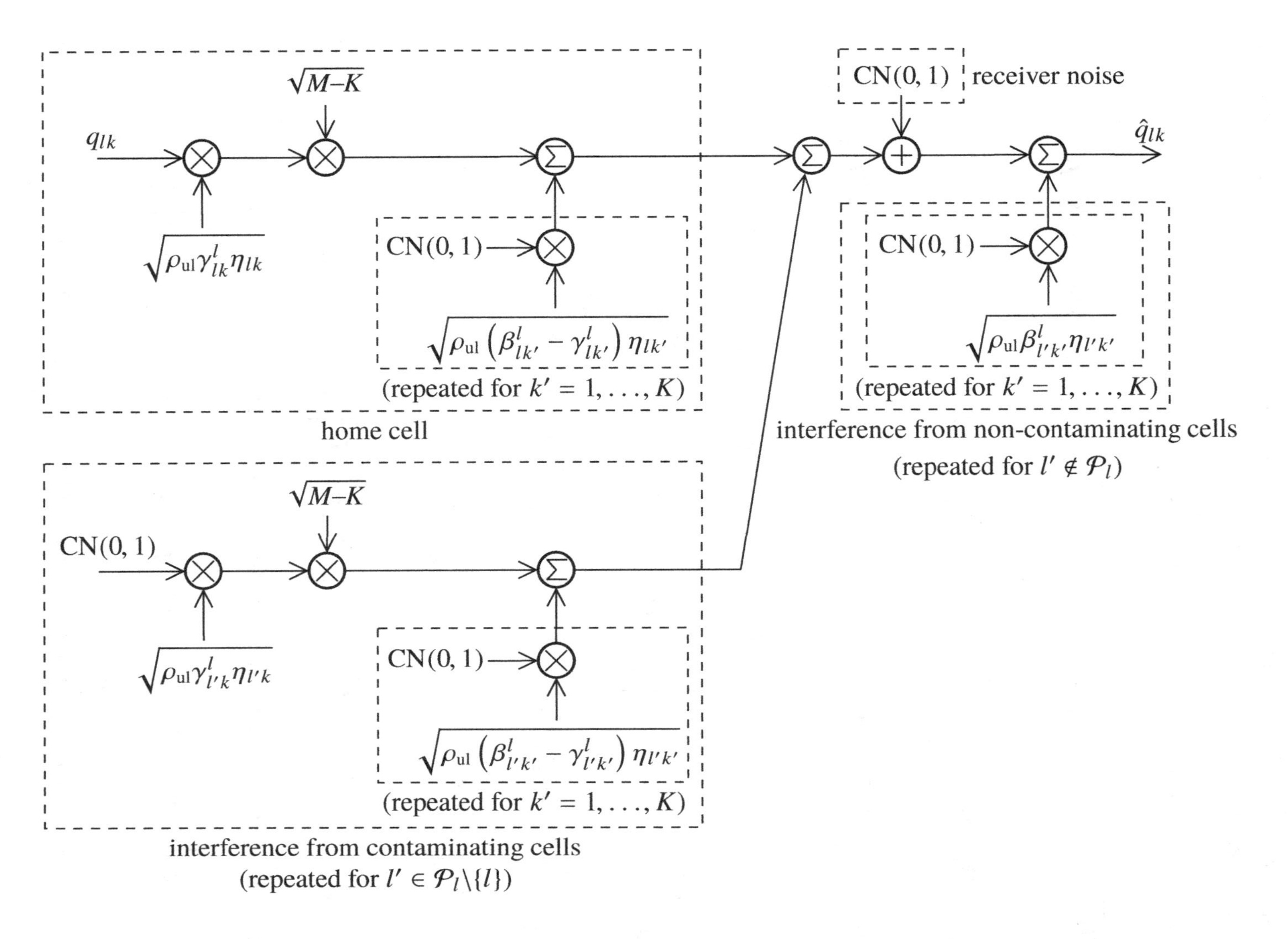

Figure 4.1. Functional block diagram for the multi-cell uplink with zero-forcing processing.

We rewrite (4.14) as the desired signal multiplied by a deterministic gain plus uncorrelated effective noise,

$$
\begin{aligned}
&\left[\frac{1}{\sqrt{M}}\boldsymbol{Z}^{l\mathrm{H}}\boldsymbol{y}_l\right]_k \\
&= \left[\sqrt{\frac{\rho_{\mathrm{ul}}}{M}}\mathsf{E}\left\{\boldsymbol{Z}^{l\mathrm{H}}\boldsymbol{Z}^l\right\}\sum_{l'\in\mathcal{P}_l}\boldsymbol{D}^{1/2}_{\gamma^l_{l'}}\boldsymbol{D}^{1/2}_{\eta_{l'}}\boldsymbol{q}_{l'}\right]_k + \left[\sqrt{\frac{\rho_{\mathrm{ul}}}{M}}\left(\boldsymbol{Z}^{l\mathrm{H}}\boldsymbol{Z}^l - \mathsf{E}\left\{\boldsymbol{Z}^{l\mathrm{H}}\boldsymbol{Z}^l\right\}\right)\sum_{l'\in\mathcal{P}_l}\boldsymbol{D}^{1/2}_{\gamma^l_{l'}}\boldsymbol{D}^{1/2}_{\eta_{l'}}\boldsymbol{q}_{l'}\right]_k \\
&+ \left[\frac{1}{\sqrt{M}}\boldsymbol{Z}^{l\mathrm{H}}\left(-\sqrt{\rho_{\mathrm{ul}}}\sum_{l'\in\mathcal{P}_l}\tilde{\boldsymbol{G}}^l_{l'}\boldsymbol{D}^{1/2}_{\eta_{l'}}\boldsymbol{q}_{l'} + \sqrt{\rho_{\mathrm{ul}}}\sum_{l'\notin\mathcal{P}_l}\boldsymbol{G}^l_{l'}\boldsymbol{D}^{1/2}_{\eta_{l'}}\boldsymbol{q}_{l'} + \boldsymbol{w}_l\right)\right]_k \\
&= \left[\sqrt{M\rho_{\mathrm{ul}}}\sum_{l'\in\mathcal{P}_l}\boldsymbol{D}^{1/2}_{\gamma^l_{l'}}\boldsymbol{D}^{1/2}_{\eta_{l'}}\boldsymbol{q}_{l'}\right]_k + \left[\sqrt{\frac{\rho_{\mathrm{ul}}}{M}}\left(\boldsymbol{Z}^{l\mathrm{H}}\boldsymbol{Z}^l - \mathsf{E}\left\{\boldsymbol{Z}^{l\mathrm{H}}\boldsymbol{Z}^l\right\}\right)\sum_{l'\in\mathcal{P}_l}\boldsymbol{D}^{1/2}_{\gamma^l_{l'}}\boldsymbol{D}^{1/2}_{\eta_{l'}}\boldsymbol{q}_{l'}\right]_k \\
&+ \left[\frac{1}{\sqrt{M}}\boldsymbol{Z}^{l\mathrm{H}}\left(-\sqrt{\rho_{\mathrm{ul}}}\sum_{l'\in\mathcal{P}_l}\tilde{\boldsymbol{G}}^l_{l'}\boldsymbol{D}^{1/2}_{\eta_{l'}}\boldsymbol{q}_{l'} + \sqrt{\rho_{\mathrm{ul}}}\sum_{l'\notin\mathcal{P}_l}\boldsymbol{G}^l_{l'}\boldsymbol{D}^{1/2}_{\eta_{l'}}\boldsymbol{q}_{l'} + \boldsymbol{w}_l\right)\right]_k \\
&= \sum_{l'\in\mathcal{P}_l}\sqrt{M\rho_{\mathrm{ul}}\gamma^l_{l'k}\eta_{l'k}}q_{l'k} + \left[\sqrt{\frac{\rho_{\mathrm{ul}}}{M}}\left(\boldsymbol{Z}^{l\mathrm{H}}\boldsymbol{Z}^l - \mathsf{E}\left\{\boldsymbol{Z}^{l\mathrm{H}}\boldsymbol{Z}^l\right\}\right)\sum_{l'\in\mathcal{P}_l}\boldsymbol{D}^{1/2}_{\gamma^l_{l'}}\boldsymbol{D}^{1/2}_{\eta_{l'}}\boldsymbol{q}_{l'}\right]_k \\
&+ \left[\frac{1}{\sqrt{M}}\boldsymbol{Z}^{l\mathrm{H}}\left(-\sqrt{\rho_{\mathrm{ul}}}\sum_{l'\in\mathcal{P}_l}\tilde{\boldsymbol{G}}^l_{l'}\boldsymbol{D}^{1/2}_{\eta_{l'}}\boldsymbol{q}_{l'} + \sqrt{\rho_{\mathrm{ul}}}\sum_{l'\notin\mathcal{P}_l}\boldsymbol{G}^l_{l'}\boldsymbol{D}^{1/2}_{\eta_{l'}}\boldsymbol{q}_{l'} + \boldsymbol{w}_l\right)\right]_k .
\end{aligned} \tag{4.15}
$$

We find the variance of the first effective noise term in (4.15) by using results in Section A.2.4,

$$
\mathsf{Var}\left\{\left[\sqrt{\frac{\rho_{\mathrm{ul}}}{M}}\left(\boldsymbol{Z}^{l\mathrm{H}}\boldsymbol{Z}^l - \mathsf{E}\left\{\boldsymbol{Z}^{l\mathrm{H}}\boldsymbol{Z}^l\right\}\right)\sum_{l'\in\mathcal{P}_l}\boldsymbol{D}^{1/2}_{\gamma^l_{l'}}\boldsymbol{D}^{1/2}_{\eta_{l'}}\boldsymbol{q}_{l'}\right]_k\right\} = \rho_{\mathrm{ul}}\sum_{l'\in\mathcal{P}_l}\sum_{k'=1}^{K}\gamma^l_{l'k'}\eta_{l'k'}. \tag{4.16}
$$

We use (4.9) to evaluate the variance of the second effective noise term in (4.15),

$$
\begin{aligned}
&\mathsf{Var}\left\{\left[\frac{1}{\sqrt{M}}\boldsymbol{Z}^{l\mathrm{H}}\left(-\sqrt{\rho_{\mathrm{ul}}}\sum_{l'\in\mathcal{P}_l}\tilde{\boldsymbol{G}}^l_{l'}\boldsymbol{D}^{1/2}_{\eta_{l'}}\boldsymbol{q}_{l'} + \sqrt{\rho_{\mathrm{ul}}}\sum_{l'\notin\mathcal{P}_l}\boldsymbol{G}^l_{l'}\boldsymbol{D}^{1/2}_{\eta_{l'}}\boldsymbol{q}_{l'} + \boldsymbol{w}_l\right)\right]_k\right\} \\
&\quad = \rho_{\mathrm{ul}}\sum_{l'\in\mathcal{P}_l}\sum_{k'=1}^{K}\left(\beta^l_{l'k'} - \gamma^l_{l'k'}\right)\eta_{l'k'} + \rho_{\mathrm{ul}}\sum_{l'\notin\mathcal{P}_l}\sum_{k'=1}^{K}\beta^l_{l'k'}\eta_{l'k'} + 1.
\end{aligned} \tag{4.17}
$$

The resulting effective SINR is

$$\mathrm{SINR}_{lk}^{\mathrm{mr,ul}} = \frac{M\rho_{\mathrm{ul}}\gamma_{lk}^{l}\eta_{lk}}{1+\rho_{\mathrm{ul}}\sum_{l'\in\mathcal{P}_l}\sum_{k'=1}^{K}\beta_{l'k'}^{l}\eta_{l'k'}+\rho_{\mathrm{ul}}\sum_{l'\notin\mathcal{P}_l}\sum_{k'=1}^{K}\beta_{l'k'}^{l}\eta_{l'k'}+M\rho_{\mathrm{ul}}\sum_{l'\in\mathcal{P}_l\setminus\{l\}}\gamma_{l'k}^{l}\eta_{l'k}}, \tag{4.18}$$

and the corresponding capacity bound is

$$C_{\mathrm{inst.},lk}^{\mathrm{mr,ul}} \geq \log_2\left(1+\mathrm{SINR}_{lk}^{\mathrm{mr,ul}}\right). \tag{4.19}$$

Functional Block Diagram

The functional block diagram in Figure 4.2 illustrates the action of maximum-ratio processing. In contrast to zero-forcing, see Figure 4.1, the coherent beamforming gain (the numerator of (4.18)) and coherent interference (the fourth term in the denominator) are proportional to M rather than to $M-K$. The non-coherent interference from contaminating cells (the second term in the denominator) is proportional to the mean-square channel rather than to the mean-square channel estimation error. Most significantly, the inter-cell interference from the non-contaminating cells, often the dominant impairment, is the same for both zero-forcing and maximum-ratio processing. In a multi-cell environment, the principal advantage of zero-forcing over maximum-ratio processing is normally a reduction of intra-cell interference. The reduced coherent interference from contaminating cells is of no consequence because the coherent beamforming gain is reduced by the same factor.

4.3 Downlink Data Transmission

Downlink data transmission entails the l'th base station independently transmitting a signal,

$$\boldsymbol{x}_{l'} = \boldsymbol{A}_{l'}\boldsymbol{D}_{\boldsymbol{\eta}_{l'}}^{1/2}\boldsymbol{q}_{l'}, \tag{4.20}$$

where $\boldsymbol{A}_{l'}$ is the precoding matrix, $\boldsymbol{\eta}_{l'}$ is a vector of the power control coefficients $\{\eta_{l'k}\}$, and $\boldsymbol{q}_{l'}$ is a vector of K symbols intended for the K terminals in that cell. The power control coefficients are non-negative and satisfy $\sum_{k=1}^{K}\eta_{l'k} \leq 1$ for all l'. The precoding matrix, $\boldsymbol{A}_{l'}$, depends only on the channel estimate within the respective cell and is normalized such that

$$\mathsf{E}\left\{\|\boldsymbol{x}_{l'}\|^2\right\} = \sum_{k=1}^{K}\eta_{l'k}. \tag{4.21}$$

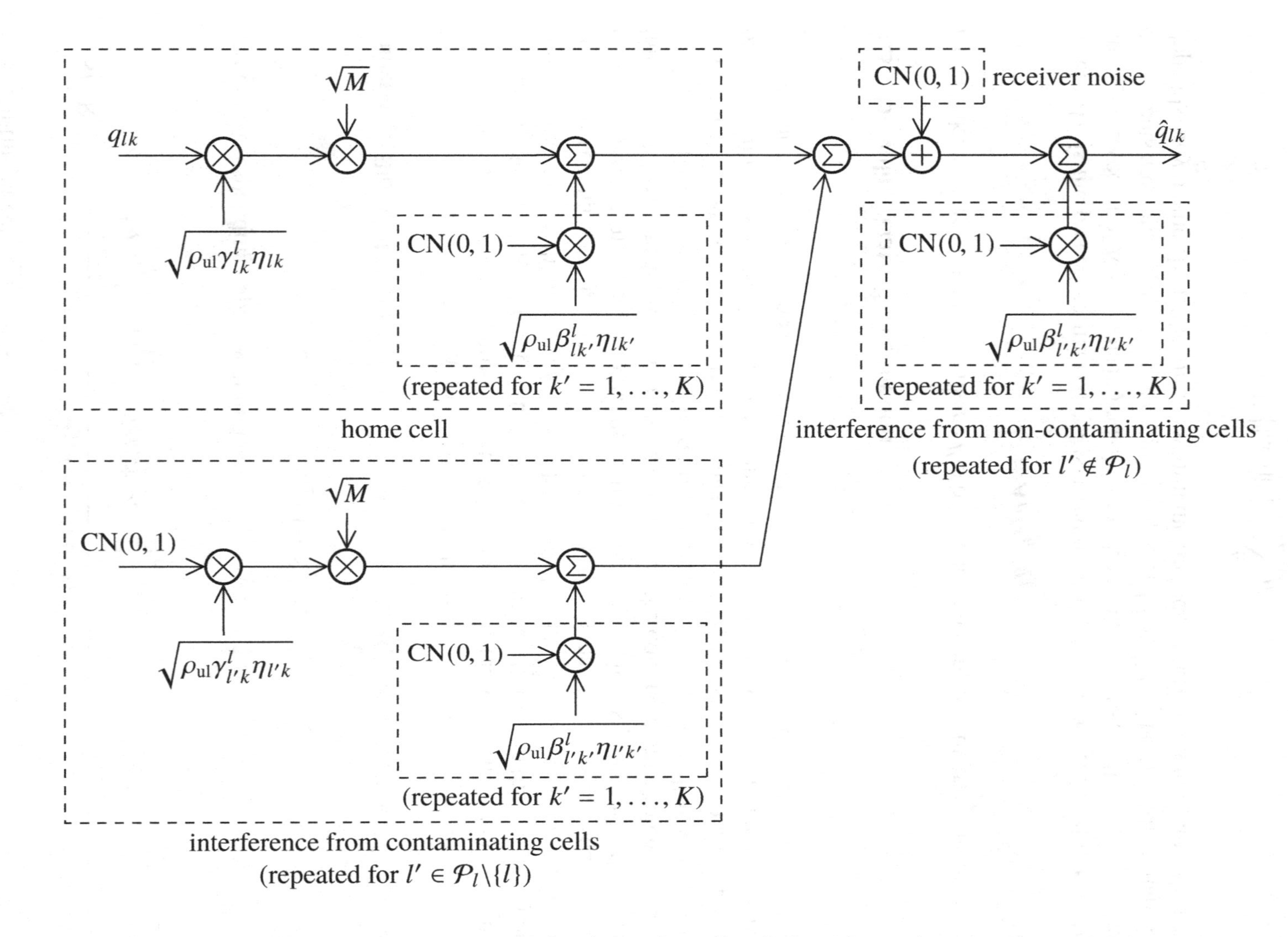

Figure 4.2. Functional block diagram for the multi-cell uplink with maximum-ratio processing.

The terminals in the lth (home) cell receive the signal,

$$\begin{aligned} \boldsymbol{y}_l &= \sqrt{\rho_{\rm dl}} \sum_{l' \in \mathcal{P}_l} \boldsymbol{G}_l^{l'{\rm T}} \boldsymbol{x}_{l'} + \sqrt{\rho_{\rm dl}} \sum_{l' \notin \mathcal{P}_l} \boldsymbol{G}_l^{l'{\rm T}} \boldsymbol{x}_{l'} + \boldsymbol{w}_l \\ &= \sqrt{\rho_{\rm dl}} \sum_{l' \in \mathcal{P}_l} \hat{\boldsymbol{G}}_l^{l'{\rm T}} \boldsymbol{x}_{l'} - \sqrt{\rho_{\rm dl}} \sum_{l' \in \mathcal{P}_l} \tilde{\boldsymbol{G}}_l^{l'{\rm T}} \boldsymbol{x}_{l'} + \sqrt{\rho_{\rm dl}} \sum_{l' \notin \mathcal{P}_l} \boldsymbol{G}_l^{l'{\rm T}} \boldsymbol{x}_{l'} + \boldsymbol{w}_l \\ &= \sqrt{\rho_{\rm dl}} \sum_{l' \in \mathcal{P}_l} \hat{\boldsymbol{G}}_l^{l'{\rm T}} \boldsymbol{A}_{l'} \boldsymbol{D}_{\eta_{l'}}^{1/2} \boldsymbol{q}_{l'} - \sqrt{\rho_{\rm dl}} \sum_{l' \in \mathcal{P}_l} \tilde{\boldsymbol{G}}_l^{l'{\rm T}} \boldsymbol{x}_{l'} + \sqrt{\rho_{\rm dl}} \sum_{l' \notin \mathcal{P}_l} \boldsymbol{G}_l^{l'{\rm T}} \boldsymbol{x}_{l'} + \boldsymbol{w}_l. \end{aligned} \tag{4.22}$$

The kth terminal receives the signal,

$$y_{lk} = \sqrt{\rho_{\rm dl}} \sum_{l' \in \mathcal{P}_l} \hat{\boldsymbol{g}}_{lk}^{l'{\rm T}} \boldsymbol{A}_{l'} \boldsymbol{D}_{\eta_{l'}}^{1/2} \boldsymbol{q}_{l'} - \sqrt{\rho_{\rm dl}} \sum_{l' \in \mathcal{P}_l} \tilde{\boldsymbol{g}}_{lk}^{l'{\rm T}} \boldsymbol{x}_{l'} + \sqrt{\rho_{\rm dl}} \sum_{l' \notin \mathcal{P}_l} \boldsymbol{g}_{lk}^{l'{\rm T}} \boldsymbol{x}_{l'} + w_{lk}. \tag{4.23}$$

All four terms in (4.23) are mutually uncorrelated. The first term contains, among others, the signal of interest and the remaining three contain only effective noise. The variance of the sum of the three effective noise terms is independent of whether zero-forcing or maximum-ratio processing is used, and is equal to

$$\begin{aligned} &\mathsf{Var}\left\{ -\sqrt{\rho_{\rm dl}} \sum_{l' \in \mathcal{P}_l} \tilde{\boldsymbol{g}}_{lk}^{l'{\rm T}} \boldsymbol{x}_{l'} + \sqrt{\rho_{\rm dl}} \sum_{l' \notin \mathcal{P}_l} \boldsymbol{g}_{lk}^{l'{\rm T}} \boldsymbol{x}_{l'} + w_{lk} \right\} \\ &\qquad = \rho_{\rm dl} \sum_{l' \in \mathcal{P}_l} \left(\beta_{lk}^{l'} - \gamma_{lk}^{l'}\right) \left(\sum_{k'=1}^{K} \eta_{l'k'} \right) + \rho_{\rm dl} \sum_{l' \notin \mathcal{P}_l} \beta_{lk}^{l'} \left(\sum_{k'=1}^{K} \eta_{l'k'} \right) + 1. \end{aligned} \tag{4.24}$$

To obtain (4.24), we have used several facts: first, the channel estimation error is independent of the channel estimates (and hence of the precoding matrices); second, the channel matrix between the home cell and the non-contaminating cells is independent of the signals transmitted by the base stations in the non-contaminating cells; and, third, the normalization (4.21).

4.3.1 Zero-Forcing

Zero-forcing uses the precoding matrix,

$$\boldsymbol{A}_{l'} = \sqrt{M-K}\,\boldsymbol{Z}^{l'*} \left(\boldsymbol{Z}^{l'{\rm T}} \boldsymbol{Z}^{l'*}\right)^{-1}. \tag{4.25}$$

The combination of (4.25) and (4.7) within the first term of (4.22) yields the following:

$$\begin{aligned}\sqrt{\rho_{\text{dl}}}\sum_{l'\in\mathcal{P}_l}\hat{\boldsymbol{G}}_l^{l'\text{T}}\boldsymbol{A}_{l'}\boldsymbol{D}_{\eta_{l'}}^{1/2}\boldsymbol{q}_{l'} &= \sqrt{\rho_{\text{dl}}}\sum_{l'\in\mathcal{P}_l}\hat{\boldsymbol{G}}_l^{l'\text{T}}\sqrt{M-K}\boldsymbol{Z}^{l'*}\left(\boldsymbol{Z}^{l'\text{T}}\boldsymbol{Z}^{l'*}\right)^{-1}\boldsymbol{D}_{\eta_{l'}}^{1/2}\boldsymbol{q}_{l'} \\ &= \sqrt{(M-K)\rho_{\text{dl}}}\sum_{l'\in\mathcal{P}_l}\boldsymbol{D}_{\gamma_l^{l'}}^{1/2}\boldsymbol{Z}^{l'\text{T}}\boldsymbol{Z}^{l'*}\left(\boldsymbol{Z}^{l'\text{T}}\boldsymbol{Z}^{l'*}\right)^{-1}\boldsymbol{D}_{\eta_{l'}}^{1/2}\boldsymbol{q}_{l'} \\ &= \sqrt{(M-K)\rho_{\text{dl}}}\sum_{l'\in\mathcal{P}_l}\boldsymbol{D}_{\gamma_l^{l'}}^{1/2}\boldsymbol{D}_{\eta_{l'}}^{1/2}\boldsymbol{q}_{l'}.\end{aligned} \tag{4.26}$$

The substitution of (4.26) into (4.23) yields the following:

$$y_{lk} = \sum_{l'\in\mathcal{P}_l}\sqrt{(M-K)\rho_{\text{dl}}\gamma_{lk}^{l'}\eta_{l'k}}q_{l'k} - \sqrt{\rho_{\text{dl}}}\sum_{l'\in\mathcal{P}_l}\tilde{\boldsymbol{g}}_{lk}^{l'\text{T}}\boldsymbol{x}_{l'} + \sqrt{\rho_{\text{dl}}}\sum_{l'\notin\mathcal{P}_l}\boldsymbol{g}_{lk}^{l'\text{T}}\boldsymbol{x}_{l'} + w_{lk}. \tag{4.27}$$

The resulting effective SINR for the kth terminal in the lth cell is

$$\text{SINR}_{lk}^{\text{zf,dl}} = \frac{(M-K)\rho_{\text{dl}}\gamma_{lk}^{l}\eta_{lk}}{1+\rho_{\text{dl}}\sum_{l'\in\mathcal{P}_l}\left(\beta_{lk}^{l'}-\gamma_{lk}^{l'}\right)\left(\sum_{k'=1}^{K}\eta_{l'k'}\right)+\rho_{\text{dl}}\sum_{l'\notin\mathcal{P}_l}\beta_{lk}^{l'}\left(\sum_{k'=1}^{K}\eta_{l'k'}\right)+(M-K)\rho_{\text{dl}}\sum_{l'\in\mathcal{P}_l\setminus\{l\}}\gamma_{lk}^{l'}\eta_{l'k}}, \tag{4.28}$$

where we have used the effective noise variance in (4.24). The corresponding capacity bound is

$$C_{\text{inst.},lk}^{\text{zf,dl}} \geq \log_2\left(1+\text{SINR}_{lk}^{\text{zf,dl}}\right). \tag{4.29}$$

Functional Block Diagram

Figure 4.3 shows the functional block diagram for the downlink with zero-forcing processing. The principal difference between downlink and uplink zero-forcing (see Figure 4.1) is that on downlink only the sum of power control coefficients matters, acting now through the kth channel only.

4.3.2 Maximum-Ratio

Maximum-ratio transmission uses the precoding matrix,

$$\boldsymbol{A}_{l'} = \frac{1}{\sqrt{M}}\boldsymbol{Z}^{l'*}. \tag{4.30}$$

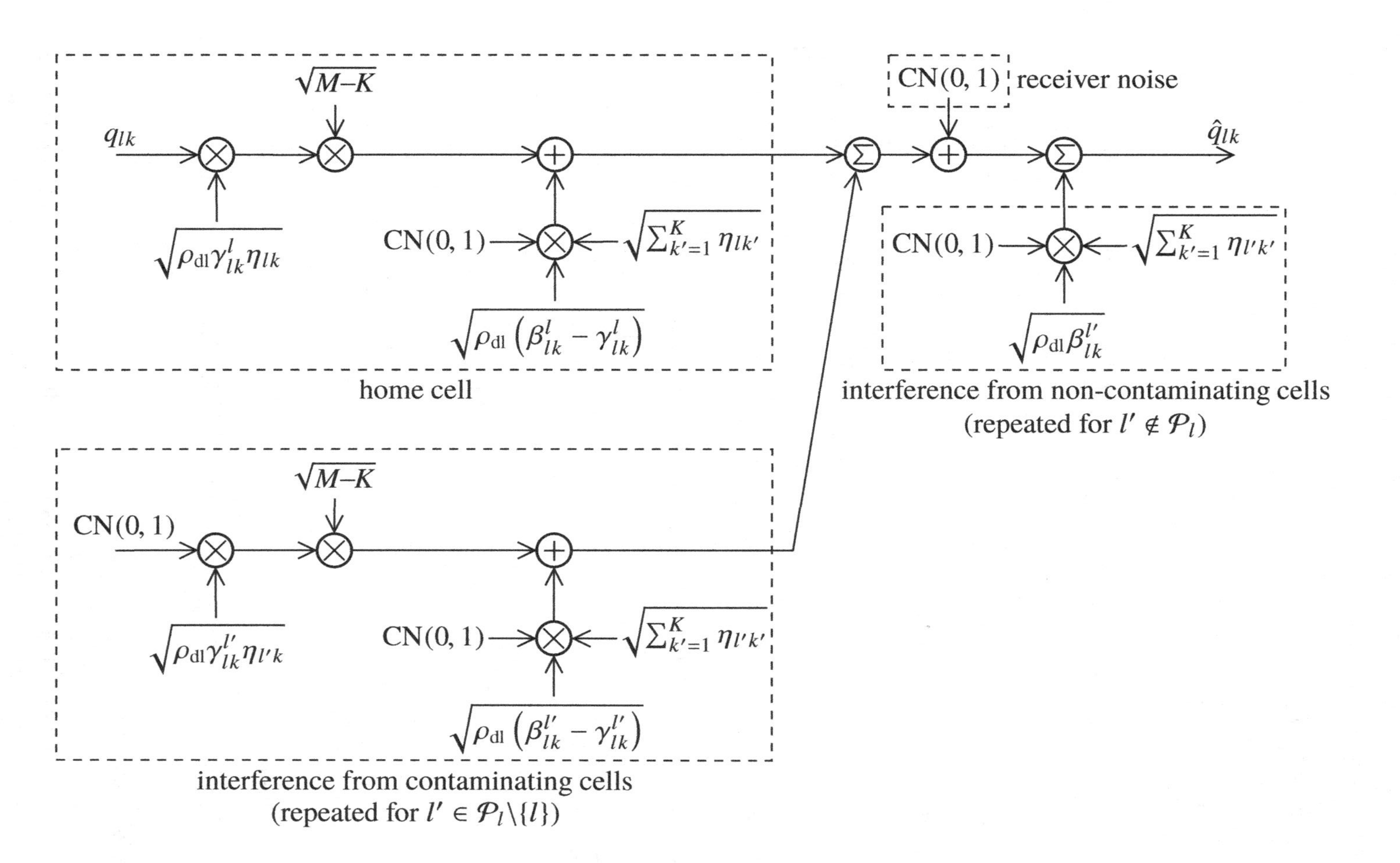

Figure 4.3. Functional block diagram for the multi-cell downlink with zero-forcing processing.

The first term in (4.22) becomes

$$
\begin{aligned}
&\sqrt{\rho_{\mathrm{dl}}} \sum_{l' \in \mathcal{P}_l} \hat{\boldsymbol{G}}_l^{l'\mathrm{T}} \boldsymbol{A}_{l'} \boldsymbol{D}_{\eta_{l'}}^{1/2} \boldsymbol{q}_{l'} \\
&= \sqrt{\frac{\rho_{\mathrm{dl}}}{M}} \sum_{l' \in \mathcal{P}_l} \boldsymbol{D}_{\gamma_l^{l'}}^{1/2} \boldsymbol{Z}^{l'\mathrm{T}} \boldsymbol{Z}^{l'*} \boldsymbol{D}_{\eta_{l'}}^{1/2} \boldsymbol{q}_{l'} \\
&= \sqrt{\frac{\rho_{\mathrm{dl}}}{M}} \sum_{l' \in \mathcal{P}_l} \boldsymbol{D}_{\gamma_l^{l'}}^{1/2} \mathsf{E}\left\{\boldsymbol{Z}^{l'\mathrm{T}} \boldsymbol{Z}^{l'*}\right\} \boldsymbol{D}_{\eta_{l'}}^{1/2} \boldsymbol{q}_{l'} + \sqrt{\frac{\rho_{\mathrm{dl}}}{M}} \sum_{l' \in \mathcal{P}_l} \boldsymbol{D}_{\gamma_l^{l'}}^{1/2} \left(\boldsymbol{Z}^{l'\mathrm{T}} \boldsymbol{Z}^{l'*} - \mathsf{E}\left\{\boldsymbol{Z}^{l'\mathrm{T}} \boldsymbol{Z}^{l'*}\right\}\right) \boldsymbol{D}_{\eta_{l'}}^{1/2} \boldsymbol{q}_{l'} \\
&= \sqrt{M \rho_{\mathrm{dl}}} \sum_{l' \in \mathcal{P}_l} \boldsymbol{D}_{\gamma_l^{l'}}^{1/2} \boldsymbol{D}_{\eta_{l'}}^{1/2} \boldsymbol{q}_{l'} + \sqrt{\frac{\rho_{\mathrm{dl}}}{M}} \sum_{l' \in \mathcal{P}_l} \boldsymbol{D}_{\gamma_l^{l'}}^{1/2} \left(\boldsymbol{Z}^{l'\mathrm{T}} \boldsymbol{Z}^{l'*} - \mathsf{E}\left\{\boldsymbol{Z}^{l'\mathrm{T}} \boldsymbol{Z}^{l'*}\right\}\right) \boldsymbol{D}_{\eta_{l'}}^{1/2} \boldsymbol{q}_{l'}. \quad (4.31)
\end{aligned}
$$

The kth terminal receives the following:

$$
\begin{aligned}
y_{lk} = &\sum_{l' \in \mathcal{P}_l} \sqrt{M \rho_{\mathrm{dl}} \gamma_{lk}^{l'} \eta_{l'k}} q_{l'k} + \left[\sqrt{\frac{\rho_{\mathrm{dl}}}{M}} \sum_{l' \in \mathcal{P}_l} \boldsymbol{D}_{\gamma_l^{l'}}^{1/2} \left(\boldsymbol{Z}^{l'\mathrm{T}} \boldsymbol{Z}^{l'*} - \mathsf{E}\left\{\boldsymbol{Z}^{l'\mathrm{T}} \boldsymbol{Z}^{l'*}\right\}\right) \boldsymbol{D}_{\eta_{l'}}^{1/2} \boldsymbol{q}_{l'}\right]_k \\
&- \sqrt{\rho_{\mathrm{dl}}} \sum_{l' \in \mathcal{P}_l} \tilde{\boldsymbol{g}}_{lk}^{l'\mathrm{T}} \boldsymbol{x}_{l'} + \sqrt{\rho_{\mathrm{dl}}} \sum_{l' \notin \mathcal{P}_l} \boldsymbol{g}_{lk}^{l'\mathrm{T}} \boldsymbol{x}_{l'} + w_{lk}. \quad (4.32)
\end{aligned}
$$

The variance of the first effective noise term in (4.32) is, similarly to (4.16),

$$
\mathsf{Var}\left\{\left[\sqrt{\frac{\rho_{\mathrm{dl}}}{M}} \sum_{l' \in \mathcal{P}_l} \boldsymbol{D}_{\gamma_l^{l'}}^{1/2} \left(\boldsymbol{Z}^{l'\mathrm{T}} \boldsymbol{Z}^{l'*} - \mathsf{E}\left\{\boldsymbol{Z}^{l'\mathrm{T}} \boldsymbol{Z}^{l'*}\right\}\right) \boldsymbol{D}_{\eta_{l'}}^{1/2} \boldsymbol{q}_{l'}\right]_k\right\} = \rho_{\mathrm{dl}} \sum_{l' \in \mathcal{P}_l} \gamma_{lk}^{l'} \left(\sum_{k'=1}^{K} \eta_{l'k'}\right). \quad (4.33)
$$

The effective SINR for the kth terminal in the lth cell is

$$
\mathrm{SINR}_{lk}^{\mathrm{mr,dl}} = \frac{M \rho_{\mathrm{dl}} \gamma_{lk}^{l} \eta_{lk}}{1 + \rho_{\mathrm{dl}} \sum_{l' \in \mathcal{P}_l} \beta_{lk}^{l'} \left(\sum_{k'=1}^{K} \eta_{l'k'}\right) + \rho_{\mathrm{dl}} \sum_{l' \notin \mathcal{P}_l} \beta_{lk}^{l'} \left(\sum_{k'=1}^{K} \eta_{l'k'}\right) + M \rho_{\mathrm{dl}} \sum_{l' \in \mathcal{P}_l \setminus \{l\}} \gamma_{lk}^{l'} \eta_{l'k}}, \quad (4.34)
$$

and the capacity bound is

$$
C_{\mathrm{inst.},lk}^{\mathrm{mr,dl}} \geq \log_2\left(1 + \mathrm{SINR}_{lk}^{\mathrm{mr,dl}}\right). \quad (4.35)
$$

Functional Block Diagram

Figure 4.4 shows the associated functional block diagram, which, compared with the corresponding diagram for zero-forcing (Figure 4.3), shows greater coherent beamforming gain but increased intra-cell interference and interference from the contaminating cells.

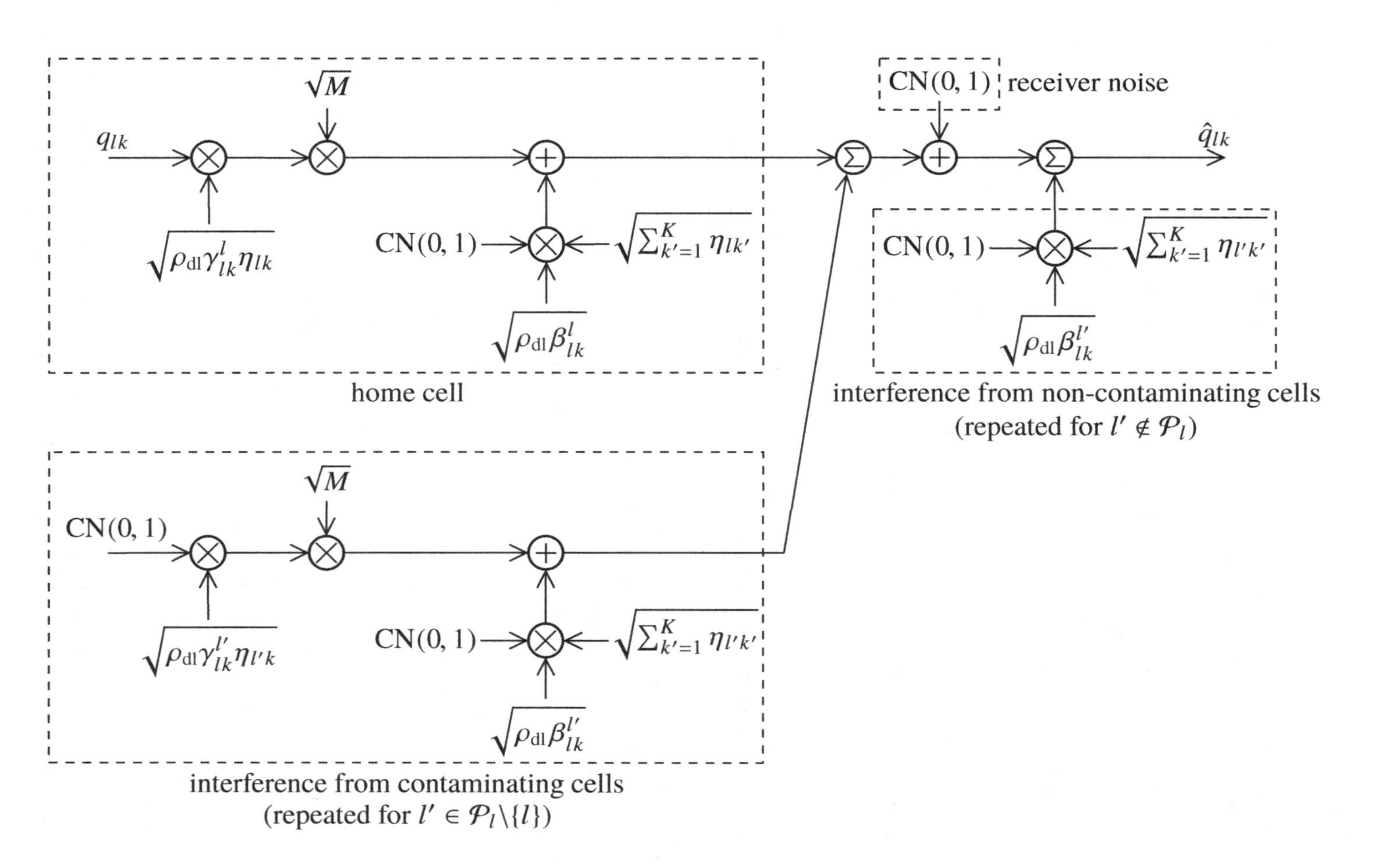

Figure 4.4. Functional block diagram for the multi-cell downlink with maximum-ratio processing.

4.4 Discussion

Table 4.1 summarizes the effective SINRs for the four multi-cell cases. These SINR expressions are remarkably similar to their single-cell counterparts in Table 3.1:

- The coherent beamforming gain term – that is, the numerator – is identical to the corresponding single-cell term, except that the mean-squared channel estimate γ_k is exchanged for γ_{lk}^l and the single-cell power control coefficients $\{\eta_k\}$ are replaced by the corresponding multi-cell coefficients $\{\eta_{lk}\}$.
- The contribution of the non-coherent interference from the home cell, contained in the second term in the denominator, is unchanged from the single-cell case.
- The power control coefficients again appear linearly both in the numerator and in the denominator.

A notable difference to the single-cell case, however, is that the effective noise now comprises two different types of interference: non-coherent and coherent.

- Non-coherent interference from contaminating cells, represented by the second term in the denominator, includes intra-cell interference in the home cell. The magnitude of this non-coherent interference for zero-forcing is proportional to the mean-square channel estimation error, while for maximum-ratio it is proportional to the mean-square channel gains.

  The magnitude of the non-coherent interference from non-contaminating cells, the third term in the denominator, is the same irrespective of whether zero-forcing or maximum-ratio is employed. This happens because transmissions from non-contaminating cells manifest themselves as uncorrelated noise in the home cell.
- Coherent interference from contaminating cells, represented by the fourth term in the denominator, grows with the number of base station antennas M at the same rate as the coherent beamforming gain does in the numerator.

4.4.1 Asymptotic Limits with Infinite Numbers of Base Station Antennas

In the limit of an infinite number of base station antennas, for the same power control coefficients, the effective SINRs are the same for zero-forcing and maximum-ratio

Uplink zero-forcing	$\dfrac{(M-K)\rho_{\mathrm{ul}}\gamma_{lk}^{l}\eta_{lk}}{1+\rho_{\mathrm{ul}}\sum\limits_{l'\in\mathcal{P}_l}\sum\limits_{k'=1}^{K}\left(\beta_{l'k'}^{l}-\gamma_{l'k'}^{l}\right)\eta_{l'k'}+\rho_{\mathrm{ul}}\sum\limits_{l'\notin\mathcal{P}_l}\sum\limits_{k'=1}^{K}\beta_{l'k'}^{l}\eta_{l'k'}+(M-K)\rho_{\mathrm{ul}}\sum\limits_{l'\in\mathcal{P}_l\setminus\{l\}}\gamma_{l'k}^{l}\eta_{l'k}}$
Uplink maximum-ratio	$\dfrac{M\rho_{\mathrm{ul}}\gamma_{lk}^{l}\eta_{lk}}{1+\rho_{\mathrm{ul}}\sum\limits_{l'\in\mathcal{P}_l}\sum\limits_{k'=1}^{K}\beta_{l'k'}^{l}\eta_{l'k'}+\rho_{\mathrm{ul}}\sum\limits_{l'\notin\mathcal{P}_l}\sum\limits_{k'=1}^{K}\beta_{l'k'}^{l}\eta_{l'k'}+M\rho_{\mathrm{ul}}\sum\limits_{l'\in\mathcal{P}_l\setminus\{l\}}\gamma_{l'k}^{l}\eta_{l'k}}$
Downlink zero-forcing	$\dfrac{(M-K)\rho_{\mathrm{dl}}\gamma_{lk}^{l}\eta_{lk}}{1+\rho_{\mathrm{dl}}\sum\limits_{l'\in\mathcal{P}_l}\left(\beta_{lk}^{l'}-\gamma_{lk}^{l'}\right)\left(\sum\limits_{k'=1}^{K}\eta_{l'k'}\right)+\rho_{\mathrm{dl}}\sum\limits_{l'\notin\mathcal{P}_l}\beta_{lk}^{l'}\left(\sum\limits_{k'=1}^{K}\eta_{l'k'}\right)+(M-K)\rho_{\mathrm{dl}}\sum\limits_{l'\in\mathcal{P}_l\setminus\{l\}}\gamma_{lk}^{l'}\eta_{l'k}}$
Downlink maximum-ratio	$\dfrac{M\rho_{\mathrm{dl}}\gamma_{lk}^{l}\eta_{lk}}{1+\rho_{\mathrm{dl}}\sum\limits_{l'\in\mathcal{P}_l}\beta_{lk}^{l'}\left(\sum\limits_{k'=1}^{K}\eta_{l'k'}\right)+\rho_{\mathrm{dl}}\sum\limits_{l'\notin\mathcal{P}_l}\beta_{lk}^{l'}\left(\sum\limits_{k'=1}^{K}\eta_{l'k'}\right)+M\rho_{\mathrm{dl}}\sum\limits_{l'\in\mathcal{P}_l\setminus\{l\}}\gamma_{lk}^{l'}\eta_{l'k}}$

Table 4.1. Effective SINR for the kth terminal in the lth cell, SINR_{lk}, in a multi-cell system.

processing,

$$\lim_{M\to\infty} \mathrm{SINR}_{lk}^{\mathrm{zf,ul}} = \lim_{M\to\infty} \mathrm{SINR}_{lk}^{\mathrm{mr,ul}}$$
$$= \frac{\gamma_{lk}^{l}\eta_{lk}}{\sum\limits_{l'\in\mathcal{P}_l\setminus\{l\}} \gamma_{l'k}^{l}\eta_{l'k}}, \tag{4.36}$$

$$\lim_{M\to\infty} \mathrm{SINR}_{lk}^{\mathrm{zf,dl}} = \lim_{M\to\infty} \mathrm{SINR}_{lk}^{\mathrm{mr,dl}}$$
$$= \frac{\gamma_{lk}^{l}\eta_{lk}}{\sum\limits_{l'\in\mathcal{P}_l\setminus\{l\}} \gamma_{lk}^{l'}\eta_{l'k}}. \tag{4.37}$$

In this limit, the performance is independent of the radiated power that is expended solely for data transmission. For particular values of the power control coefficients $\{\eta_{lk}\}$, the expressions (4.36) and (4.37) coincide with those obtained in [26]; see Appendix E.

4.4.2 The Effects of Pilot Contamination

Pilot contamination has two effects on the effective SINR: first, it reduces the magnitude of the mean-square channel estimate $\gamma_{l'k}^{l}$, and hence degrades the coherent beamforming gain in the numerator; second, it gives rise to coherent interference – the fourth term in the denominator. Unless M is very large, the effect on the coherent beamforming gain is typically the dominant impairment.

4.4.3 Non-Synchronous Pilot Interference

Throughout this chapter, we have assumed that the activities in all cells occur synchronously, that is, during the training part of each coherence interval all terminals in all cells send pilot sequences to their respective base stations. The pilot sequences used in the home cell l are reused in all cells indexed by $\mathcal{P}_l$. This reuse results in pilot contamination, which manifests itself through degradation of the coherent beamforming gain and in the presence of coherent interference. Yet, in terms of the magnitude of the pilot contamination effect, it makes no fundamental difference whether the received pilots in the home cell are affected by interference that stems from pilots synchronously transmitted by terminals in other cells, or non-synchronously transmitted pilots from other cells, or payload data from other cells. The reason is that any signal that is transmitted by terminals in other cells can always be expanded in terms of the orthogonal pilot sequences that are used in the home cell.

In more detail, consider the special case of only two cells: a home cell l and an interfering cell l', both with K terminals. If the terminals both in the home cell and in the interfering cell synchronously transmit the same uplink pilots $\boldsymbol{\Phi}_l$, as assumed in Section 4.1, then (4.2) becomes

$$\left[\boldsymbol{Y}'_{pl}\right]_{mk} = \sqrt{\tau_p \rho_{ul}} g_{lk}^{lm} + \sqrt{\tau_p \rho_{ul}} g_{l'k}^{lm} + \left[\boldsymbol{W}'_{pl}\right]_{mk}. \tag{4.38}$$

The first term in (4.38) represents the channel of interest, and the second term represents the contaminating channel response from the interfering cell. The second term has variance,

$$\tau_p \rho_{ul} \beta_{l'k}^l. \tag{4.39}$$

Now suppose that, by way of contrast, during the training phase of the home cell, the K terminals in the interfering cell instead transmit non-synchronous pilots, or random payload data, $\boldsymbol{X}_{l'}$, with uncorrelated components and at maximum permitted power. Then the counterpart to (4.1), the $M \times \tau_p$ pilot signal received by the base station in the home cell, is

$$\boldsymbol{Y}_{pl} = \sqrt{\tau_p \rho_{ul}} \boldsymbol{G}_l^l \boldsymbol{\Phi}_l^H + \sqrt{\rho_{ul}} \boldsymbol{G}_{l'}^l \boldsymbol{X}_{l'} + \boldsymbol{W}_{pl}. \tag{4.40}$$

As in Section 4.1, by projecting $\boldsymbol{Y}_{pl}$ onto $\boldsymbol{\Phi}_l$, and using the fact that $\boldsymbol{\Phi}_l^H \boldsymbol{\Phi}_l = \boldsymbol{I}_K$, we obtain the statistic $\boldsymbol{Y}'_{pl} = \boldsymbol{Y}_{pl} \boldsymbol{\Phi}_l$ that is used for the channel estimation. The (m, k)th element of $\boldsymbol{Y}'_{pl}$ is equal to

$$\left[\boldsymbol{Y}'_{pl}\right]_{mk} = \sqrt{\tau_p \rho_{ul}} g_{lk}^{lm} + \sqrt{\rho_{ul}} \left[\boldsymbol{G}_{l'}^l \boldsymbol{X}_{l'} \boldsymbol{\Phi}_l\right]_{mk} + \left[\boldsymbol{W}_{pl} \boldsymbol{\Phi}_l\right]_{mk}. \tag{4.41}$$

Equation (4.41) should be compared to (4.38). In both expressions, the second term of (4.41), $\sqrt{\rho_{ul}} \left[\boldsymbol{G}_{l'}^l \boldsymbol{X}_{l'} \boldsymbol{\Phi}_l\right]_{mk}$, represents contamination from the interfering cell. This term has zero mean and variance,

$$\rho_{ul} \sum_{k'=1}^{K} \beta_{l'k'}^l. \tag{4.42}$$

If K is of the same order as τ_p (recall that $\tau_p \geq K$ always), then the variances of the contamination terms in (4.39) and (4.42) are of the same order too. As a consequence, the effect of the pilot contamination caused by payload or non-synchronous pilot transmission in other cells is comparable to the effect of the pilot contamination caused by transmission of synchronous pilots.

4.5 Summary of Key Points

- Table 4.1 gives rigorous lower bounds on the instantaneous ergodic capacity for the kth terminal in the lth cell according to the formula $C_{\text{inst.},lk} = \log_2(1 + \text{SINR}_{lk})$. Instantaneous capacity is converted to net spectral efficiency as in Section 3.6. These bounds account for all imperfections associated with channel estimation errors, channel non-orthogonality, beamforming gain uncertainty, and pilot reuse.

  Appendix H summarizes the notation used in Table 4.1. The power control coefficients $\{\eta_{lk}\}$ satisfy $0 \leq \eta_{lk} \leq 1$ for all l and k on the uplink. In the downlink, they satisfy $\eta_{lk} \geq 0$ for all l and k, and

$$\sum_{k=1}^{K} \eta_{lk} \leq 1 \tag{4.43}$$

  for all l.

- Compared to the single-cell case, new types of effective noise appear. In total, the effective noise comprises: (i) non-coherent interference from contaminating cells, including the home cell; (ii) non-coherent interference from non-contaminating cells; and (iii) coherent interference from contaminating cells, but excluding the home cell. The power of these effective noises are given by the second, third, and fourth terms of the denominator of the effective SINR expressions in Table 4.1.

- The functional block diagrams in Figures 4.1–4.4 provide interpretations of the different terms that constitute the effective SINR: coherent beamforming gain of the desired signal, non-coherent interference, coherent interference, and receiver noise.

- While we have assumed throughout the chapter that the number of terminals in each cell is equal to K, the results can easily be generalized to the case of a different number of terminals per cell at the expense of some additional notation.

Chapter 5

POWER CONTROL PRINCIPLES

Effective and computationally tractable power control is one of the unique new features of Massive MIMO. Among other things, power control handles near-far effects, and it enables uniformly good service throughout the cell. Massive MIMO power control occurs on a long time scale because effective SINRs depend only on large-scale fading coefficients. In this chapter, we develop power control schemes to meet given performance targets both in single-cell and multi-cell systems and both for the uplink and the downlink, with max-min (egalitarian) SINR fairness as a particularly important special case.

5.1 Preliminaries

An inspection of effective SINRs in Table 3.1 and Table 4.1 discloses qualitatively identical dependence on the power control coefficients for the four cases of uplink/downlink and zero-forcing/maximum-ratio. This permits a unified treatment of power control.

In the single-cell case, from Table 3.1 we observe that the effective SINR for terminal k can always be written in the following general form:

$$\text{SINR}_k = \frac{a_k \eta_k}{1 + \sum\limits_{k'=1}^{K} b_k^{k'} \eta_{k'}}, \tag{5.1}$$

where $\{a_k\}$ and $\{b_k^{k'}\}$ are strictly positive constants, given by Table 5.1. In Table 5.1, M, K, ρ_{ul}, ρ_{dl}, β_k, γ_k and η_k have the meanings as defined in Chapter 3, and summarized in Appendix H.

	Zero-Forcing	Maximum-Ratio
Uplink	$a_k = (M-K)\rho_{\mathrm{ul}}\gamma_k$ $b_k^{k'} = \rho_{\mathrm{ul}}\left(\beta_{k'} - \gamma_{k'}\right)$	$a_k = M\rho_{\mathrm{ul}}\gamma_k$ $b_k^{k'} = \rho_{\mathrm{ul}}\beta_{k'}$
Downlink	$a_k = (M-K)\rho_{\mathrm{dl}}\gamma_k$ $b_k^{k'} = \rho_{\mathrm{dl}}\left(\beta_k - \gamma_k\right)$	$a_k = M\rho_{\mathrm{dl}}\gamma_k$ $b_k^{k'} = \rho_{\mathrm{dl}}\beta_k$

Table 5.1. Explicit formulas for the coefficients $\{a_k\}$ and $\{b_k^{k'}\}$ in (5.1) for a single-cell system.

	Zero-Forcing	Maximum-Ratio
Uplink	$a_{lk} = (M-K)\rho_{\mathrm{ul}}\gamma_{lk}^l$ $b_{lk}^{l'k'} = \rho_{\mathrm{ul}}\left(\beta_{l'k'}^l - \gamma_{l'k'}^l\right)$ $c_{lk}^{l'k'} = \rho_{\mathrm{ul}}\beta_{l'k'}^l$ $d_{lk}^{l'} = (M-K)\rho_{\mathrm{ul}}\gamma_{l'k}^l$	$a_{lk} = M\rho_{\mathrm{ul}}\gamma_{lk}^l$ $b_{lk}^{l'k'} = \rho_{\mathrm{ul}}\beta_{l'k'}^l$ $c_{lk}^{l'k'} = \rho_{\mathrm{ul}}\beta_{l'k'}^l$ $d_{lk}^{l'} = M\rho_{\mathrm{ul}}\gamma_{l'k}^l$
Downlink	$a_{lk} = (M-K)\rho_{\mathrm{dl}}\gamma_{lk}^l$ $b_{lk}^{l'k'} = \rho_{\mathrm{dl}}\left(\beta_{lk}^{l'} - \gamma_{lk}^{l'}\right)$ $c_{lk}^{l'k'} = \rho_{\mathrm{dl}}\beta_{lk}^{l'}$ $d_{lk}^{l'} = (M-K)\rho_{\mathrm{dl}}\gamma_{lk}^{l'}$	$a_{lk} = M\rho_{\mathrm{dl}}\gamma_{lk}^l$ $b_{lk}^{l'k'} = \rho_{\mathrm{dl}}\beta_{lk}^{l'}$ $c_{lk}^{l'k'} = \rho_{\mathrm{dl}}\beta_{lk}^{l'}$ $d_{lk}^{l'} = M\rho_{\mathrm{dl}}\gamma_{lk}^{l'}$

Table 5.2. Explicit formulas for the coefficients $\{a_{lk}\}$, $\{b_{lk}^{l'k'}\}$, $\{c_{lk}^{l'k'}\}$ and $\{d_{lk}^{l'}\}$ in (5.2) for a multi-cell system.

	Single-Cell	Multi-Cell
Uplink	$0 \leq \eta_k \leq 1$ $k = 1, \ldots, K$	$0 \leq \eta_{lk} \leq 1$ $k = 1, \ldots, K, \quad l = 1, \ldots, L$
Downlink	$\sum_{k=1}^{K} \eta_k \leq 1$ and $\eta_k \geq 0, \quad k = 1, \ldots, K$	$\sum_{k=1}^{K} \eta_{lk} \leq 1, \quad l = 1, \ldots, L$ and $\eta_{lk} \geq 0, \quad k = 1, \ldots, K, \quad l = 1, \ldots, L$

Table 5.3. Summary of constraints on the power control coefficients.

Similarly, for the multi-cell case, from Table 4.1 the effective SINR for the kth terminal in the lth cell can be written as

$$\mathrm{SINR}_{lk} = \frac{a_{lk}\eta_{lk}}{1 + \sum_{l' \in \mathcal{P}_l} \sum_{k'=1}^{K} b_{lk}^{l'k'} \eta_{l'k'} + \sum_{l' \notin \mathcal{P}_l} \sum_{k'=1}^{K} c_{lk}^{l'k'} \eta_{l'k'} + \sum_{l' \in \mathcal{P}_l \setminus \{l\}} d_{lk}^{l'} \eta_{l'k}}, \tag{5.2}$$

where the non-negative coefficients $\{a_{lk}\}$, $\{b_{lk}^{l'k'}\}$, $\{c_{lk}^{l'k'}\}$, and $\{d_{lk}^{l'}\}$ are given in Table 5.2. In Table 5.2, $\beta_{l'k'}^{l}$, $\gamma_{l'k}^{l}$ and η_{lk} are defined in Chapter 4; see Appendix H for a summary of their meanings. The single-cell scenario is, of course, a special case of the multi-cell scenario, obtained by setting $\{c_{lk}^{l'k'}\}$ and $\{d_{lk}^{l'}\}$ equal to zero and omitting the cell index l.

Table 5.3 summarizes the constraints on the power control coefficients for the single-cell and multi-cell cases. Henceforth, we denote by L the total number of cells.

5.2 Power Control with Given SINR Targets

We show next that the problem of designing a power control policy that offers guaranteed quality-of-service can be cast as a linear feasibility problem. The key observation is that the numerator and denominator of (5.1) and (5.2) are each linear in the power control coefficients.

5.2.1 Single-Cell System

We start with a single-cell system. Consider a constraint of the form

$$\text{SINR}_k \geq \overline{\text{SINR}}_k, \qquad k = 1, \ldots, K, \tag{5.3}$$

where $\overline{\text{SINR}}_k$ is a given *target SINR* for the kth terminal. An SINR target is directly translatable to a spectral efficiency target, by using the formulas for net spectral efficiency in Section 3.6. In practice, such a target could reflect a quality-of-service requirement for a particular terminal. The set of constraints in (5.3) is equivalent to the following set of inequalities:

$$a_k \eta_k \geq \overline{\text{SINR}}_k \left(1 + \sum_{k'=1}^{K} b_k^{k'} \eta_{k'} \right), \qquad k = 1, \ldots, K, \tag{5.4}$$

which are linear in $\{\eta_k\}$. This means that the problem of designing a power control policy under which the kth terminal achieves an SINR of at least $\overline{\text{SINR}}_k$ can be written as

$$\begin{aligned} &\text{find} \quad \{\eta_k\} \\ &\text{subject to} \quad \text{(i)} \ \ \text{SINR}_k \geq \overline{\text{SINR}}_k, \qquad k = 1, \ldots, K, \\ &\qquad\qquad\quad \text{(ii)} \ \text{the constraints in Table 5.3.} \end{aligned} \tag{5.5}$$

Problem (5.5) is a linear programming feasibility problem, which is easily solved using standard software toolboxes. The set of all SINR constraints in (5.3) can be satisfied for some permissible $\{\eta_k\}$ if and only if the problem (5.5) has a solution.

5.2.2 Multi-Cell System

For the multi-cell case, we again impose a target SINR as a set of constraints,

$$\text{SINR}_{lk} \geq \overline{\text{SINR}}_{lk}, \qquad k = 1, \ldots, K, \qquad l = 1, \ldots, L \tag{5.6}$$

where $\overline{\text{SINR}}_{lk}$ is a target SINR for the kth terminal in the lth cell. Each inequality in (5.6) is equivalent to the following inequality:

$$a_{lk} \eta_{lk} \geq \overline{\text{SINR}}_{lk} \left(1 + \sum_{l' \in \mathcal{P}_l} \sum_{k'=1}^{K} b_{lk}^{l'k'} \eta_{l'k'} + \sum_{l' \notin \mathcal{P}_l} \sum_{k'=1}^{K} c_{lk}^{l'k'} \eta_{l'k'} + \sum_{l' \in \mathcal{P}_l \setminus \{l\}} d_{lk}^{l'} \eta_{l'k} \right), \tag{5.7}$$

which is linear in $\{\eta_{lk}\}$. Hence, a power control policy design problem of the form

$$\begin{aligned} \text{find} \quad & \{\eta_{lk}\} \\ \text{subject to} \quad & \text{(i)} \ \text{SINR}_{lk} \geq \overline{\text{SINR}}_{lk}, \quad k = 1, \ldots, K, \quad l = 1, \ldots, L \\ & \text{(ii) the constraints in Table 5.3} \end{aligned} \tag{5.8}$$

is a linear programming feasibility problem, as in the single-cell case.

5.3 Max-Min Fairness Power Control

An important design philosophy for power control policies is max-min (egalitarian) fairness, which seeks to maximize the worst SINR over all terminals. A simple proof by contradiction establishes that the max-min solution to the optimization problem provides equal SINRs for all terminals. Assume the contrary; then there is a terminal whose SINR is greater than the max-min SINR. We can reduce the power control coefficient for that terminal somewhat, which can only affect the other terminals by reducing their denominators, thereby increasing their SINRs. Consequently, the original assumption of a max-min solution is false. Max-min fairness power control therefore amounts to setting all SINR targets equal to a common value $\overline{\text{SINR}}$, and then finding the largest possible value of $\overline{\text{SINR}}$ that ensures that all constraints in Table 5.3 are satisfied.

For a single-cell system, max-min fairness means that the SINR targets for all terminals in the cell are equal. In a multi-cell system, max-min fairness may be imposed network-wide, or independently within each cell. We discuss these different possibilities in more detail next.

5.3.1 Single-Cell System with Max-Min Fairness

First, consider a single-cell system. Setting the SINR targets of all terminals in the cell equal to a common value $\overline{\text{SINR}}$ amounts to requiring that

$$\overline{\text{SINR}}_k = \overline{\text{SINR}}, \qquad k = 1, \ldots, K. \tag{5.9}$$

Explicitly, the max-min philosophy then results in the following optimization problem:

$$\begin{aligned} \text{maximize} \quad & \overline{\text{SINR}} \\ \text{with respect to} \quad & \{\eta_k\} \\ \text{subject to} \quad & \text{(i)} \ \text{SINR}_k \geq \overline{\text{SINR}}, \quad k = 1, \ldots, K \\ & \text{(ii) the constraints in Table 5.3.} \end{aligned} \tag{5.10}$$

All inequalities involved in (5.10) are linear and hence (5.10) is a quasi-linear programming problem. Such a problem can in general be solved by performing a bisection search over $\overline{\text{SINR}}$ and for each candidate value of $\overline{\text{SINR}}$, solving a linear feasibility problem. However, for the specific problem (5.10) simple closed-form solutions exist. We summarize these solutions in Table 5.4, and give their derivations in what follows.

Uplink

Consider first the uplink. It is clear from (5.1) that for both zero-forcing and maximum-ratio processing at least one of the coefficients $\{\eta_k\}$ must be equal to unity. To see why, suppose this is not the case so that $\eta_k < 1$ for $k = 1, \ldots, K$. Then all $\{\eta_k\}$ could be scaled by a common constant such that at least one of them becomes equal to unity. This scaling would increase all values $\{\text{SINR}_k\}$, which contradicts the supposed optimality of the solution of the original $\{\eta_k\}$. Hence, at the optimum it must hold that

$$\begin{aligned} \text{SINR}_k &= \overline{\text{SINR}}, \qquad k = 1, \ldots, K, \qquad \text{for some } \overline{\text{SINR}}, \\ \eta_k &= 1, \qquad \text{for at least one } k, \end{aligned} \tag{5.11}$$

where $\overline{\text{SINR}}$ is the optimal common SINR. From (5.1) and (5.11), we have that

$$a_k \eta_k = \overline{\text{SINR}} \left(1 + \sum_{k'=1}^{K} b_k^{k'} \eta_{k'} \right), \qquad k = 1, \ldots, K. \tag{5.12}$$

For the uplink case, $b_k^{k'}$ has no dependence on k, so the right-hand side of (5.12) is a constant with respect to k. Since $\{\eta_k\}$ satisfy $0 \leq \eta_k \leq 1$ for all k, and $\eta_k = 1$ for some k, we must have that

$$\eta_k = \frac{\min_{k'} \{a_{k'}\}}{a_k}. \tag{5.13}$$

The resulting $\overline{\text{SINR}}$ is found by inserting (5.13) into (5.1), which yields

$$\overline{\text{SINR}} = \frac{1}{\frac{1}{\min_{k'} \{a_{k'}\}} + \sum_{k'=1}^{K} \frac{b_k^{k'}}{a_{k'}}}, \tag{5.14}$$

independently of k. The substitution of the expressions for $\{a_k\}$ and $\{b_k^{k'}\}$ from Table 5.1 yields the formulas listed in Table 5.4.

	Zero-Forcing	Maximum-Ratio
Uplink	$\eta_k = \frac{\min_{k'} \{\gamma_{k'}\}}{\gamma_k}$ $\overline{\mathrm{SINR}} = \frac{(M-K)\rho_{\mathrm{ul}}}{\frac{1}{\min_k \{\gamma_k\}} + \rho_{\mathrm{ul}} \sum_{k=1}^{K} \frac{\beta_k - \gamma_k}{\gamma_k}}$	$\eta_k = \frac{\min_{k'} \{\gamma_{k'}\}}{\gamma_k}$ $\overline{\mathrm{SINR}} = \frac{M\rho_{\mathrm{ul}}}{\frac{1}{\min_k \{\gamma_k\}} + \rho_{\mathrm{ul}} \sum_{k=1}^{K} \frac{\beta_k}{\gamma_k}}$
Downlink	$\eta_k = \frac{1 + \rho_{\mathrm{dl}}(\beta_k - \gamma_k)}{\rho_{\mathrm{dl}}\gamma_k \left(\frac{1}{\rho_{\mathrm{dl}}} \sum_{k'=1}^{K} \frac{1}{\gamma_{k'}} + \sum_{k'=1}^{K} \frac{\beta_{k'} - \gamma_{k'}}{\gamma_{k'}} \right)}$ $\overline{\mathrm{SINR}} = \frac{(M-K)\rho_{\mathrm{dl}}}{\sum_{k=1}^{K} \frac{1}{\gamma_k} + \rho_{\mathrm{dl}} \sum_{k=1}^{K} \frac{\beta_k - \gamma_k}{\gamma_k}}$	$\eta_k = \frac{1 + \rho_{\mathrm{dl}}\beta_k}{\rho_{\mathrm{dl}}\gamma_k \left(\frac{1}{\rho_{\mathrm{dl}}} \sum_{k'=1}^{K} \frac{1}{\gamma_{k'}} + \sum_{k'=1}^{K} \frac{\beta_{k'}}{\gamma_{k'}} \right)}$ $\overline{\mathrm{SINR}} = \frac{M\rho_{\mathrm{dl}}}{\sum_{k=1}^{K} \frac{1}{\gamma_k} + \rho_{\mathrm{dl}} \sum_{k=1}^{K} \frac{\beta_k}{\gamma_k}}$

Table 5.4. Power control coefficients and resulting common SINR values, $\overline{\mathrm{SINR}}$, for max-min fairness power control in a single-cell system.

Downlink

For the downlink, we infer from (5.1) that a max-min solution requires that the power constraint $\sum_{k=1}^{K} \eta_k = 1$ be satisfied with equality. To see why this is so, suppose $\sum_{k=1}^{K} \eta_k < 1$. Then, since $b_k^{k'} > 0$, by scaling all values $\{\eta_k\}$ by a common scaling factor such that $\sum_{k=1}^{K} \eta_k$ increases, SINR_k would increase for all k, which contradicts the max-min optimality of the solution. Hence, at the optimum we must have that

$$\text{SINR}_k = \overline{\text{SINR}}, \qquad k = 1, \ldots, K, \qquad \text{for some } \overline{\text{SINR}}, \tag{5.15}$$

$$\sum_{k=1}^{K} \eta_k = 1, \tag{5.16}$$

where $\overline{\text{SINR}}$ is the max-min optimal common SINR. The combination of (5.15) and (5.1) yields

$$a_k \eta_k = \overline{\text{SINR}} \left(1 + \sum_{k'=1}^{K} b_k^{k'} \eta_{k'} \right), \qquad k = 1, \ldots, K. \tag{5.17}$$

Here in the downlink case, $b_k^{k'}$ does not depend on k' so we let $b_k = b_k^{k'}$. Hence, by using (5.16) in (5.17), we obtain

$$\eta_k = \frac{\overline{\text{SINR}}(1 + b_k)}{a_k}. \tag{5.18}$$

Using (5.16) again, we conclude that

$$\begin{aligned} \overline{\text{SINR}} \sum_{k=1}^{K} \frac{1 + b_k}{a_k} &= \sum_{k=1}^{K} \eta_k \\ &= 1. \end{aligned} \tag{5.19}$$

Hence

$$\overline{\text{SINR}} = \frac{1}{\sum_{k=1}^{K} \frac{1 + b_k}{a_k}}, \tag{5.20}$$

and

$$\eta_k = \frac{1 + b_k}{a_k \sum_{k''=1}^{K} \frac{1 + b_{k''}}{a_{k''}}}. \tag{5.21}$$

The substitution of the expressions for $\{a_k\}$ and $\{b_k\}$ from Table 5.1 into (5.20) and (5.21) gives the formulas in Table 5.4.

The Effect of Adding Extra Terminals

One effect of max-min power control is that comparatively little power is expended on terminals that enjoy strong channels. It is clear from Table 5.4 that η_k decreases as γ_k (and therefore β_k) increases. We further investigate this phenomenon by adding a new terminal with a stronger channel than the existing terminals to a cell that is already giving max-min service to K terminals. An inspection of the denominators of the max-min SINRs in Table 5.4 discloses that the addition of a new (stronger) terminal is most disruptive when all $K+1$ values of $\{\beta_k\}$ (and therefore $\{\gamma_k\}$) are equal, in which case the denominator increases at most by a factor of $(K+1)/K$. For maximum-ratio, the numerator is unaffected by the new terminal, and for zero-forcing the numerator decreases by the factor $(M-K-1)/(M-K)$. The general conclusion is that when K is large most power is spent on serving terminals that are subject to severe large-scale fading, and service to additional terminals close to the base station can be given almost for free.

Zero-Forcing versus Maximum-Ratio Processing

It is natural to ask when zero-forcing is preferable to maximum-ratio processing. The results of Table 5.4 provide a remarkably simple and explicit answer: for both uplink and downlink, $\overline{\text{SINR}}^{\text{zf}} > \overline{\text{SINR}}^{\text{mr}}$ if and only if $\overline{\text{SINR}}^{\text{mr}} > 1$. To prove this result for the uplink, assume that $\overline{\text{SINR}}^{\text{zf,ul}} > \overline{\text{SINR}}^{\text{mr,ul}}$, substitute the two expressions for $\overline{\text{SINR}}^{\text{zf,ul}}$ and $\overline{\text{SINR}}^{\text{mr,ul}}$ from Table 5.4 into the inequality, and simplify. The equivalent inequality is that

$$\frac{1}{\min_k \{\gamma_k\}} + \rho_{\text{ul}} \sum_{k=1}^{K} \frac{\beta_k}{\gamma_k} < M\rho_{\text{ul}}, \tag{5.22}$$

which holds precisely when $\overline{\text{SINR}}^{\text{mr,ul}} > 1$. A similar calculation establishes the desired result for the downlink.

5.3.2 Multi-Cell Systems with Network-Wide Max-Min Fairness

For a multi-cell system with network-wide max-min fairness power control, we set all target SINRs equal:

$$\overline{\text{SINR}}_{lk} = \overline{\text{SINR}}, \qquad k = 1, \ldots, K, \qquad l = 1, \ldots, L. \tag{5.23}$$

This results in the following optimization problem:

$$\begin{array}{rll} \text{maximize} & \overline{\text{SINR}} & \\ \text{with respect to} & \{\eta_{lk}\} & \\ \text{subject to} & \text{(i)}\ \ \text{SINR}_{lk} \geq \overline{\text{SINR}}, \quad k = 1, \ldots, K, \quad l = 1, \ldots, L & \\ & \text{(ii)}\ \ \text{the constraints in Table 5.3.} & \end{array} \tag{5.24}$$

All inequalities involved in (5.24) are linear and hence (5.24) is a quasi-linear programming problem. Modern software toolboxes can solve this problem even for large L. However, since (5.24) yields power control coefficients such that all terminals in all cells obtain the same SINR, the power control coefficients in a given cell, say the lth cell, will depend on the conditions in other cells, $l' \neq l$, that are far away.

Specifically, suppose some cell in the network has a low throughput, due to overcrowding with terminals, or because a particular terminal that is scheduled for service experiences severe shadow fading. Then that low throughput is unnecessarily imposed on all terminals served in all other cells. In particular, the value of $\overline{\text{SINR}}$ achieved in (5.24) may approach zero as $L \to \infty$. To see why, consider first the common SINR achieved with max-min power control in the single-cell case, given in Table 5.4. Clearly,

$$\begin{aligned} \overline{\text{SINR}}^{\text{ul}} &\leq M\rho_{\text{ul}} \min_k \{\gamma_k\} \\ &\leq M\rho_{\text{ul}} \min_k \{\beta_k\} \end{aligned} \tag{5.25}$$

$$\begin{aligned} \overline{\text{SINR}}^{\text{dl}} &\leq M\rho_{\text{dl}} \min_k \{\gamma_k\} \\ &\leq M\rho_{\text{dl}} \min_k \{\beta_k\}, \end{aligned} \tag{5.26}$$

irrespective of whether zero-forcing or maximum-ratio processing is used. Next, note that an upper bound on $\overline{\text{SINR}}$ achieved in (5.24) will be given by the corresponding single-cell $\overline{\text{SINR}}$ in Table 5.4 for the most disadvantaged cell in the network, the cell with the smallest per-terminal throughput. Hence, the optimal $\overline{\text{SINR}}$ achieved in (5.24) cannot exceed $M\rho_{\text{ul}} \min_{l,k}\{\beta_{lk}^l\}$ in the uplink, and $M\rho_{\text{dl}} \min_{l,k}\{\beta_{lk}^l\}$ in the downlink. In lognormal shadow fading, $\min_{l,k}\{\beta_{lk}^l\} \to 0$ as $L \to \infty$. This renders network-wide max-min fairness power control fundamentally unscalable with respect to the number of cells, L.

5.3.3 Per-Cell Power Control for Negligible Coherent Interference and Full Power

A remedy to the scalability problem of network-wide max-min fairness power control is to equalize the SINRs only within each cell. Next, we give an algorithm that does this, in the special case when the coherent interference is negligible and all cells use the maximum allowed power. Specifically, let us make the following two assumptions:

- The coherent interference, that is the fourth term in the denominator of (5.2), is negligible,

$$d_{lk}^{l'} = 0, \tag{5.27}$$

  for all l, l' and k. Excluding the coherent interference term does not mean that we are ignoring pilot contamination; recall that for moderate values of M, the major effect of pilot contamination is to reduce the coherent gain through a reduction in $\{\gamma_{lk}^{l'}\}$.

  Under the assumption (5.27), (5.2) simplifies as follows:

$$\mathrm{SINR}_{lk} = \frac{a_{lk}\eta_{lk}}{1 + \sum\limits_{l' \in \mathcal{P}_l} \sum\limits_{k'=1}^{K} b_{lk}^{l'k'} \eta_{l'k'} + \sum\limits_{l' \notin \mathcal{P}_l} \sum\limits_{k'=1}^{K} c_{lk}^{l'k'} \eta_{l'k'}}. \tag{5.28}$$

- Each cell uses the full available power; that is, in the uplink, at least one terminal in each cell transmits with maximum power,

$$\eta_{lk} = 1, \qquad \text{for some } k \text{ for every } l = 1, \ldots, L \tag{5.29}$$

  and in the downlink, all base stations expend the maximum available power,

$$\sum_{k=1}^{K} \eta_{lk} = 1, \qquad l = 1, \ldots, L. \tag{5.30}$$

Under the assumptions stated, max-min power control can be performed within each cell independently. At the resulting operating point, all terminals in each cell achieve a common cell-specific SINR value, $\overline{\mathrm{SINR}}_l$. Tables 5.5 and 5.6 summarize the results, and derivations are given in what follows. Naturally, the single-cell results in Table 5.4 are special cases of the multi-cell results in Tables 5.5 and 5.6.

With the power control strategy described here, each cell in the network is equally important and no cell dictates what other cells should do. By contrast, with the network-wide equal throughput strategy (see Section 5.3.2), the throughput is determined by the most disadvantaged cell.

Zero-Forcing	$\eta_{lk} = \frac{\min_{k'} \left\{\gamma^l_{lk'}\right\}}{\gamma^l_{lk}}$ $\overline{\text{SINR}}_l = \frac{(M-K)\rho_{\text{ul}}}{\frac{1}{\min_k \left\{\gamma^l_{lk}\right\}} + \rho_{\text{ul}} \sum_{l' \in \mathcal{P}_l} \frac{\min_k \left\{\gamma^{l'}_{l'k}\right\}}{\min_k \left\{\gamma^l_{lk}\right\}} \sum_{k=1}^{K} \frac{\beta^l_{l'k} - \gamma^l_{l'k}}{\gamma^{l'}_{l'k}} + \rho_{\text{ul}} \sum_{l' \notin \mathcal{P}_l} \frac{\min_k \left\{\gamma^{l'}_{l'k}\right\}}{\min_k \left\{\gamma^l_{lk}\right\}} \sum_{k=1}^{K} \frac{\beta^l_{l'k}}{\gamma^{l'}_{l'k}}}$
Maximum-Ratio	$\eta_{lk} = \frac{\min_{k'} \left\{\gamma^l_{lk'}\right\}}{\gamma^l_{lk}}$ $\overline{\text{SINR}}_l = \frac{M\rho_{\text{ul}}}{\frac{1}{\min_k \left\{\gamma^l_{lk}\right\}} + \rho_{\text{ul}} \sum_{l' \in \mathcal{P}_l} \frac{\min_k \left\{\gamma^{l'}_{l'k}\right\}}{\min_k \left\{\gamma^l_{lk}\right\}} \sum_{k=1}^{K} \frac{\beta^l_{l'k}}{\gamma^{l'}_{l'k}} + \rho_{\text{ul}} \sum_{l' \notin \mathcal{P}_l} \frac{\min_k \left\{\gamma^{l'}_{l'k}\right\}}{\min_k \left\{\gamma^l_{lk}\right\}} \sum_{k=1}^{K} \frac{\beta^l_{l'k}}{\gamma^{l'}_{l'k}}}$

Table 5.5. Uplink power control coefficients, and per-cell max-min SINR, $\overline{\text{SINR}}_l$, when coherent interference is negligible and "full power" is utilized in each cell. (The equations for maximum-ratio processing may be simplified by combining the sums over $l' \in \mathcal{P}_l$ and $l' \notin \mathcal{P}_l$, but we have not done that here in order to maintain the symmetry of exposition with the zero-forcing case.)

Zero-Forcing	$\eta_{lk} = \dfrac{1 + \sum_{l' \in \mathcal{P}_l} \rho_{\text{dl}} \left(\beta_{lk}^{l'} - \gamma_{lk}^{l'}\right) + \sum_{l' \notin \mathcal{P}_l} \rho_{\text{dl}} \beta_{lk}^{l'}}{\gamma_{lk}^{l} \left(\sum_{k'=1}^{K} \frac{1}{\gamma_{lk'}^{l}} + \rho_{\text{dl}} \sum_{l' \in \mathcal{P}_l} \sum_{k'=1}^{K} \frac{\beta_{lk'}^{l'} - \gamma_{lk'}^{l'}}{\gamma_{lk'}^{l}} + \rho_{\text{dl}} \sum_{l' \notin \mathcal{P}_l} \sum_{k'=1}^{K} \frac{\beta_{lk'}^{l'}}{\gamma_{lk'}^{l}} \right)}$ $\overline{\text{SINR}}_l = \dfrac{(M-K)\rho_{\text{dl}}}{\sum_{k=1}^{K} \frac{1}{\gamma_{lk}^{l}} + \rho_{\text{dl}} \sum_{l' \in \mathcal{P}_l} \sum_{k=1}^{K} \frac{\beta_{lk}^{l'} - \gamma_{lk}^{l'}}{\gamma_{lk}^{l}} + \rho_{\text{dl}} \sum_{l' \notin \mathcal{P}_l} \sum_{k=1}^{K} \frac{\beta_{lk}^{l'}}{\gamma_{lk}^{l}}}$
Maximum-Ratio	$\eta_{lk} = \dfrac{1 + \sum_{l' \in \mathcal{P}_l} \rho_{\text{dl}} \beta_{lk}^{l'} + \sum_{l' \notin \mathcal{P}_l} \rho_{\text{dl}} \beta_{lk}^{l'}}{\gamma_{lk}^{l} \left(\sum_{k'=1}^{K} \frac{1}{\gamma_{lk'}^{l}} + \rho_{\text{dl}} \sum_{l' \in \mathcal{P}_l} \sum_{k'=1}^{K} \frac{\beta_{lk'}^{l'}}{\gamma_{lk'}^{l}} + \rho_{\text{dl}} \sum_{l' \notin \mathcal{P}_l} \sum_{k'=1}^{K} \frac{\beta_{lk'}^{l'}}{\gamma_{lk'}^{l}} \right)}$ $\overline{\text{SINR}}_l = \dfrac{M\rho_{\text{dl}}}{\sum_{k=1}^{K} \frac{1}{\gamma_{lk}^{l}} + \rho_{\text{dl}} \sum_{l' \in \mathcal{P}_l} \sum_{k=1}^{K} \frac{\beta_{lk}^{l'}}{\gamma_{lk}^{l}} + \rho_{\text{dl}} \sum_{l' \notin \mathcal{P}_l} \sum_{k=1}^{K} \frac{\beta_{lk}^{l'}}{\gamma_{lk}^{l}}}$

Table 5.6. Downlink power control coefficients, and per-cell max-min SINR, $\overline{\text{SINR}}_l$, when coherent interference is negligible and "full power" is utilized in each cell. (As in Table 5.5, the equations for maximum-ratio processing may be simplified by combining terms.)

Uplink

In the uplink, $\{b_{lk}^{l'k'}\}$ and $\{c_{lk}^{l'k'}\}$ in (5.28) are independent of k. If we denote $b_l^{l'k'} = b_{lk}^{l'k'}$ and $c_l^{l'k'} = c_{lk}^{l'k'}$, then (5.28) becomes

$$\mathrm{SINR}_{lk} = \frac{a_{lk}\eta_{lk}}{1 + \sum_{l' \in \mathcal{P}_l} \sum_{k'=1}^{K} b_l^{l'k'} \eta_{l'k'} + \sum_{l' \notin \mathcal{P}_l} \sum_{k'=1}^{K} c_l^{l'k'} \eta_{l'k'}}. \tag{5.31}$$

In (5.31), the denominator is independent of k, and SINR_{lk} has the same form as in the single-cell case. Since every cell uses full power in the sense of (5.29), the techniques for the single-cell case (see Section 5.3.1) can be applied. In particular, the following choice of power control coefficients yield nominal max-min fairness in the lth cell, for all l,

$$\eta_{lk} = \frac{\min_{k'} \{a_{lk'}\}}{a_{lk}}, \qquad k = 1, \ldots, K, \qquad l = 1, \ldots, L, \tag{5.32}$$

and the SINR achieved by all terminals in the lth cell is

$$\begin{aligned} \overline{\mathrm{SINR}}_l &= \frac{\min_{k'} \{a_{lk'}\}}{1 + \sum_{l' \in \mathcal{P}_l} \sum_{k'=1}^{K} b_l^{l'k'} \eta_{l'k'} + \sum_{l' \notin \mathcal{P}_l} \sum_{k'=1}^{K} c_l^{l'k'} \eta_{l'k'}} \\ &= \frac{1}{\frac{1}{\min_{k'} \{a_{lk'}\}} + \sum_{l' \in \mathcal{P}_l} \frac{\min_{k'} \{a_{l'k'}\}}{\min_{k'} \{a_{lk'}\}} \sum_{k'=1}^{K} \frac{b_l^{l'k'}}{a_{l'k'}} + \sum_{l' \notin \mathcal{P}_l} \frac{\min_{k'} \{a_{l'k'}\}}{\min_{k'} \{a_{lk'}\}} \sum_{k'=1}^{K} \frac{c_l^{l'k'}}{a_{l'k'}}}, \end{aligned} \tag{5.33}$$

independent of k. Insertion of the formulas from Table 5.2 into (5.32) and (5.33) yields the results in Table 5.5.

Downlink

In the downlink, $\{b_{lk}^{l'k'}\}$ and $\{c_{lk}^{l'k'}\}$ do not depend on k', so we can write $b_{lk}^{l'} = b_{lk}^{l'k'}$ and $c_{lk}^{l'} = c_{lk}^{l'k'}$. With this new notation, and under the assumption (5.30), (5.28) becomes

$$
\begin{aligned}
\text{SINR}_{lk} &= \frac{a_{lk}\eta_{lk}}{1 + \sum_{l' \in \mathcal{P}_l} \sum_{k'=1}^{K} b_{lk}^{l'} \eta_{l'k'} + \sum_{l' \notin \mathcal{P}_l} \sum_{k'=1}^{K} c_{lk}^{l'} \eta_{l'k'}} \\
&= \frac{a_{lk}\eta_{lk}}{1 + \sum_{l' \in \mathcal{P}_l} b_{lk}^{l'} \sum_{k'=1}^{K} \eta_{l'k'} + \sum_{l' \notin \mathcal{P}_l} c_{lk}^{l'} \sum_{k'=1}^{K} \eta_{l'k'}} \\
&= \frac{a_{lk}\eta_{lk}}{1 + \sum_{l' \in \mathcal{P}_l} b_{lk}^{l'} + \sum_{l' \notin \mathcal{P}_l} c_{lk}^{l'}},
\end{aligned}
\tag{5.34}
$$

which is of the same form as in the single-cell case; there is no dependence on k' in the denominator. Using a similar argument as in Section 5.3.1, we find that the following power control coefficients yield max-min optimality in each cell:

$$
\eta_{lk} = \frac{1 + \sum_{l' \in \mathcal{P}_l} b_{lk}^{l'} + \sum_{l' \notin \mathcal{P}_l} c_{lk}^{l'}}{a_{lk} \sum_{k''=1}^{K} \frac{1 + \sum_{l' \in \mathcal{P}_l} b_{lk''}^{l'} + \sum_{l' \notin \mathcal{P}_l} c_{lk''}^{l'}}{a_{lk''}}}.
\tag{5.35}
$$

The resulting SINR nominally achieved by all terminals in the lth cell is

$$
\overline{\text{SINR}}_l = \frac{1}{\sum_{k''=1}^{K} \frac{1 + \sum_{l' \in \mathcal{P}_l} b_{lk''}^{l'} + \sum_{l' \notin \mathcal{P}_l} c_{lk''}^{l'}}{a_{lk''}}},
\tag{5.36}
$$

independent of k. Insertion of the expressions in Table 5.2 into (5.35) and (5.36) yields the results in Table 5.6.

Discussion

While the power control conditions (5.29) and (5.30) would generally be considered reasonable cellular practice, the assumption of negligible coherent interference (5.27) may be violated in some scenarios. In Section 6.2.5, we give a heuristic power control algorithm that accounts for significant coherent interference.

5.3.4 Uniformly Good Service

The application of max-min fairness power control ensures that all terminals enjoy uniformly good service. However, in practice, because of path loss and shadow fading, some terminals are likely to have a very small $\beta^{l}_{l'k}$. With max-min power control in its pure form as described above, substantial resources would be allocated to make sure that these terminals are well served. This in turn may impose a significant penalty on the throughput of all others. It is therefore often prudent to drop a small fraction of the terminals from service before computing the power control coefficients. In the case studies presented in Chapter 6, we will exclude a small percentage of the terminals from service in any given cell.

Alternatively, instead of dropping disadvantaged terminals from service entirely, one could give them some minimal SINR, which would manifest itself as additional linear constraints on the power control coefficients. Likewise, users demanding, or willing to pay for, extra service could be allocated a much higher guaranteed SINR than what the typical terminal receives.

5.4 Summary of Key Points

- In Massive MIMO, the power control coefficients $\{\eta_k\}$ (for single-cell systems) and $\{\eta_{lk}\}$ (for multi-cell systems) depend only on the large-scale fading coefficients $\{\beta_k\}$ respectively $\{\beta^{l}_{l'k}\}$.

- For single-cell systems, Table 5.4 gives explicit formulas for power control coefficients $\{\eta_k\}$ that yield uniformly good throughputs in the cell in the max-min fairness sense. The resulting $\{\mathrm{SINR}_k\}$ for all terminals are equal to a common value $\overline{\mathrm{SINR}}$, which is as large as possible under the given power constraints.

- For multi-cell systems, a network-wide solution to the max-min fairness SINR optimization problem, that equalizes the SINRs of all terminals in all cells, can be obtained by solving a quasi-linear optimization problem as described in Section 5.3.2.

  In case the coherent interference can be neglected, and each cell uses full power in the sense made precise by (5.29) and (5.30), then power control can be performed such that max-min fairness holds within each cell; that is, all terminals in the lth cell achieve a common SINR, $\overline{\mathrm{SINR}}_l$, where $\overline{\mathrm{SINR}}_l$ may fluctuate from cell to cell. Tables 5.5 and 5.6 give the power control coefficients $\{\eta_{lk}\}$ that achieve this, along with the resulting values of $\{\overline{\mathrm{SINR}}_l\}$.

- With max-min fairness power control, a small fraction of the terminals in each cell may be dropped from service prior to computing the power control coefficients in order to ensure that no deeply shadowed terminal dictates an unnecessarily low throughput on all others.

Chapter 6

CASE STUDIES

The case studies in this chapter are of two types: first, a single isolated cell for rural broadband fixed access (Section 6.1); second, multi-cell deployments for dense urban and suburban mobile access (Section 6.3). We model all important physical phenomena, including randomness of terminal locations, path loss, and shadow fading, and use the capacity expressions derived in Chapters 3–5. These expressions account for the effects of intra- and inter-cell interference, channel estimation errors, and the cost of pilot transmission. While all capacity bounds in Chapters 3 and 4 are rigorous and all algorithms in Chapter 5 provide exact solutions to precise optimization problems, in the multi-cell design examples some heuristic algorithms are needed for terminal-to-base station assignment, pilot assignment, and power control; we describe these algorithms in Section 6.2.

Tables 6.1–6.3 summarize all parameters used in the three design examples, and the resulting performance. The numbers given in Tables 6.2 and 6.3 represent 95% likely values (over the randomness associated with the large-scale fading), for the coverage probabilities specified in Table 6.1. Specifically, in the rural scenario, all 3000 homes obtain 20 Mb/s in the downlink and 10 Mb/s in the uplink, i.e., the coverage probability is 100%; Tables 6.2 and 6.3 list the numbers of antennas that are needed, with 95% probability, to offer this service. In the mobile access scenario, the coverage probability is 95% – that is, 5% of all terminals are dropped from service. The throughput numbers in Tables 6.2 and 6.3 represent the 95% likely throughput for the terminals that remain in service. For the mobile access scenarios, therefore, the overall reliability is equal to 0.95×0.95.

6.1 Single-Cell Deployment Example: Fixed Broadband Access in Rural Area

A single Massive MIMO base station serves 3000 homes in a rural area with data rates comparable to cable- or fiber-based access. We assume an isolated cell, for example a rural

	Rural fixed access	Dense urban mobile access	Suburban mobile access
Carrier frequency	800 MHz	1.9 GHz	1.9 GHz
Spectral bandwidth	20 MHz	20 MHz	20 MHz
Cell radius	11.3 km	500 m	2 km
Average number of terminals per cell	3000	18	18
Coverage probability	100%	95%	95%
Base station antenna gain	0 dBi	0 dBi	0 dBi
Terminal antenna gain	6 dBi	0 dBi	0 dBi
Base station receiver noise figure	9 dB	9 dB	9 dB
Terminal receiver noise figure	9 dB	9 dB	9 dB
Nominal noise temperature	300 K	300 K	300 K
Terminal mobility	Stationary	142 km/h[a]	284 km/h[a]
Coherence time	50 ms	2 ms	1 ms
Coherence bandwidth	300 kHz	210 kHz	210 kHz
Shadow fading standard deviation	8 dB	8 dB	8 dB
Shadow fading diversity	best of two	none	none
Path loss model	Hata	COST 231	COST 231
Base station antenna height	32 m	30 m	30 m
Terminal antenna height	5 m	1.5 m	1.5 m
Uplink pilot reuse factor	N/A	7	3
Total radiated power per base station	10 W	1 W	1 W
Radiated power per terminal	1 W	200 mW	200 mW

[a]With a factor-of-two design margin (see Section 2.1.4), these velocities are instead 71 km/h respectively 142 km/h.

Table 6.1. Summary of parameters used in the Massive MIMO case studies.

	Rural fixed access	Dense urban mobile access	Suburban mobile access
Downlink net throughput per terminal	20 Mb/s	4.5 Mb/s	3.1 Mb/s
Uplink net throughput per terminal	10 Mb/s	2.8 Mb/s	1.1 Mb/s
Number of base station antennas	3 200	64	256

Table 6.2. Summary of performance with zero-forcing processing. All numbers in the table represent 95% likely performance figures.

	Rural fixed access	Dense urban mobile access	Suburban mobile access
Downlink net throughput per terminal	20 Mb/s	4.8 Mb/s	3.2 Mb/s
Uplink net throughput per terminal	10 Mb/s	3.2 Mb/s	1.1 Mb/s
Number of base station antennas	8 200	64	256

Table 6.3. Summary of performance with maximum-ratio processing. All numbers in the table represent 95% likely performance figures.

town, which is therefore free from inter-cell interference.

The system operates in the 800 MHz band with a total spectral bandwidth of 20 MHz. The net throughput requirement in the uplink is 10 Mb/s for each subscriber, and 20 Mb/s in the downlink. The max-min SINR power control scheme in Section 5.3.1, summarized in Table 5.4, ensures that every subscriber enjoys uniformly good service.

The base station array is located 32 m above the ground. Each home has a terminal with a fixed 6 dBi gain antenna, mounted outdoors 5 m above the ground. The Hata model [31] determines the path loss. The shadow fading for the terminal is the best of two independent realizations of a log-normal random variable having 0 dB mean and 8 dB standard deviation. This simulates the choice of one of two possible locations on the exterior of the house for the installation of the terminal antenna. We assume that the large-scale fading coefficients for each house are constant over time.

The stationary nature of the channel permits a relatively long coherence time of $T_c = 50$ ms. At 800 MHz carrier frequency, the coherence time may in fact be much longer since the terminals are stationary, but we use a more conservative value to cap the latency between uplink and downlink. Also changes in the environment, vehicle motion and trees swaying in the wind, induce fluctuations in the channel. We furthermore assume a coherence bandwidth of $B_c = 300$ kHz. Thus, the sample duration of the coherence interval is $\tau_c = B_c T_c = 15000$.

We split the coherence interval between the uplink and the downlink in proportion to the throughput requirements. That is, after excluding the samples expended on uplink pilots, $1/3$ of each coherence interval is used for uplink data transmission and $2/3$ is used for downlink data transmission.

We limit the uplink pilot overhead, somewhat arbitrarily, to 20%, and assign each terminal a unique orthogonal pilot sequence, so we can serve $0.2 \times \tau_c = 3000$ homes. With a deployment density of 7.5 homes/km^2, a cell that contains 3000 homes has the radius,

$$\sqrt{\frac{3000}{\pi \cdot 7.5}} \approx 11.3 \text{ km}. \tag{6.1}$$

6.1.1 Required Number of Antennas and Radiated Power

The pilot overhead leaves 80% of the coherence interval for data transmission. To obtain net throughput, we multiply all instantaneous spectral efficiencies by $(1/3) \times 0.8$ for the uplink, and $(2/3) \times 0.8$ for the downlink, to account for the resource split between uplink and downlink, and for pilot overhead. The overhead associated with the possible need for a cyclic prefix was neglected here.

The required instantaneous rate per terminal in both uplink and downlink is therefore 37.5 Mb/s, which is equivalent to an instantaneous spectral efficiency of 1.875 b/s/Hz. Consequently, the required SINR is $2^{1.875} - 1 = 2.668$ (4.26 dB). It remains to determine the number of antennas, M, and the radiated power, required to achieve this SINR. We first compute how many base station antennas are needed to achieve the target uplink throughput, and subsequently find the downlink radiated power required to achieve the downlink target throughput.

Figure 6.1 shows the cumulative distribution of the number of antennas, M, that are needed to achieve the uplink target throughput of 10 Mb/s/terminal for different uplink radiated powers. We obtained this figure by generating 5000 independent random realizations of the terminal locations and shadow fading profiles. To obtain the normalized SNRs, ρ_{ul} and ρ_{dl}, from absolute transmit powers, noise temperatures, and noise figures, we used a standard link budget calculation; see Appendix F.

The superiority of zero-forcing relative to maximum-ratio processing for this scenario is apparent: at 1 W uplink power, and 95% reliability, zero-forcing requires only $M = 3200$ while maximum-ratio requires $M = 8200$. Increasing the radiated power per terminal beyond 1 W has only a minor effect; with 2, 4, and 8 W uplink power, the required number of antennas is 3100, 3050, and 3030 for zero-forcing processing, and 8100, 8050, and 8030 for maximum-ratio processing. On the other hand, reducing the uplink power below 1 W

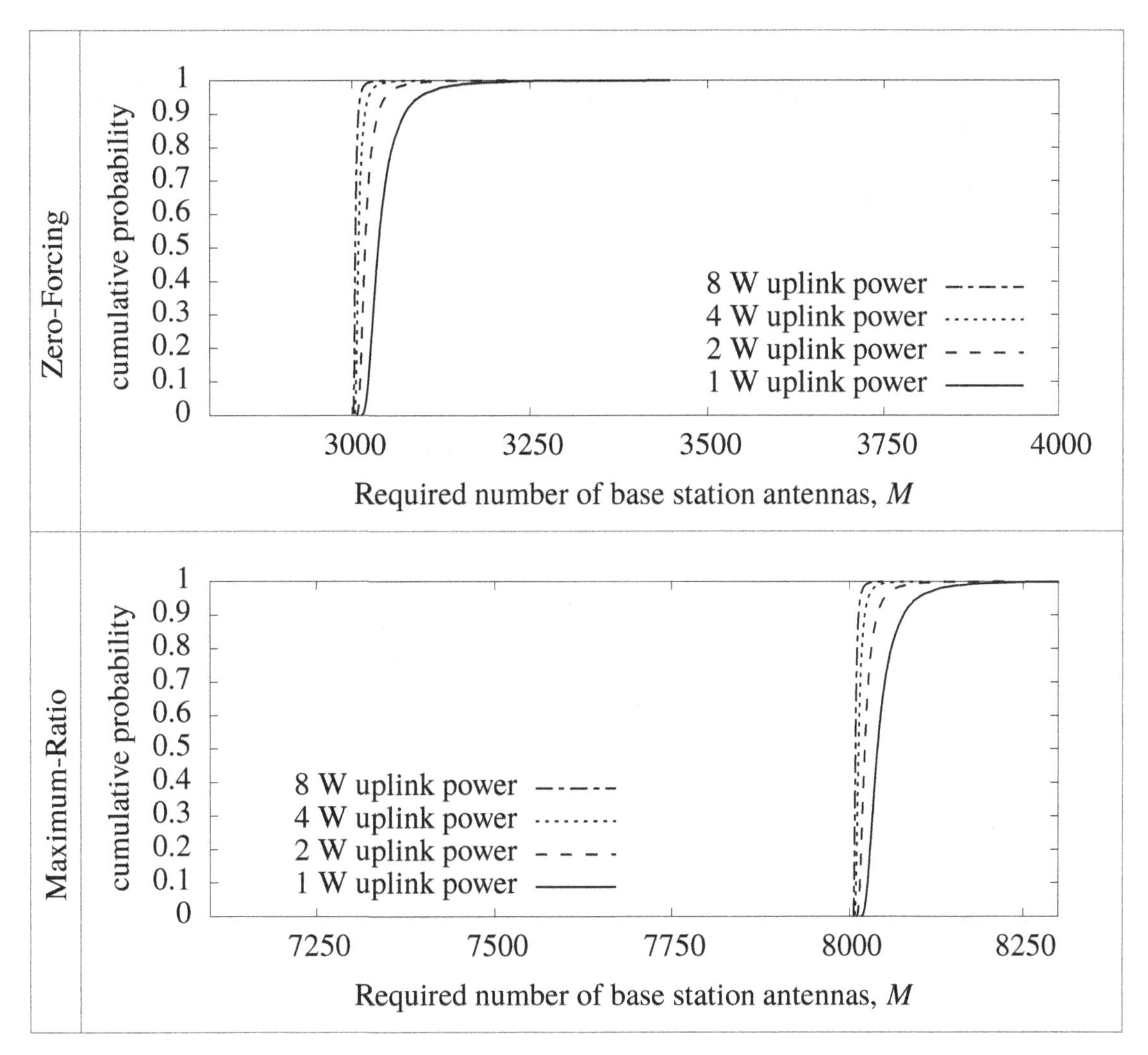

Figure 6.1. Single-cell example: Required number of base station antennas, M, for zero-forcing and maximum-ratio processing, for a range of uplink powers.

would notably increase the required number of base station antennas (not shown in the figure). To understand the phenomenology, three points should be kept in mind:

- The use of zero-forcing requires that $M > K$; so irrespective of power levels, the number of antennas must be greater than 3000.
- As shown in Section 3.4.3, for maximum-ratio processing the SINR per terminal can be no greater than M/K. Hence, given the stipulated SINR, the minimum number of antennas is $3000 \times 2.668 = 8004$.
- The uplink power affects both the quality of the channel estimates and the noisiness

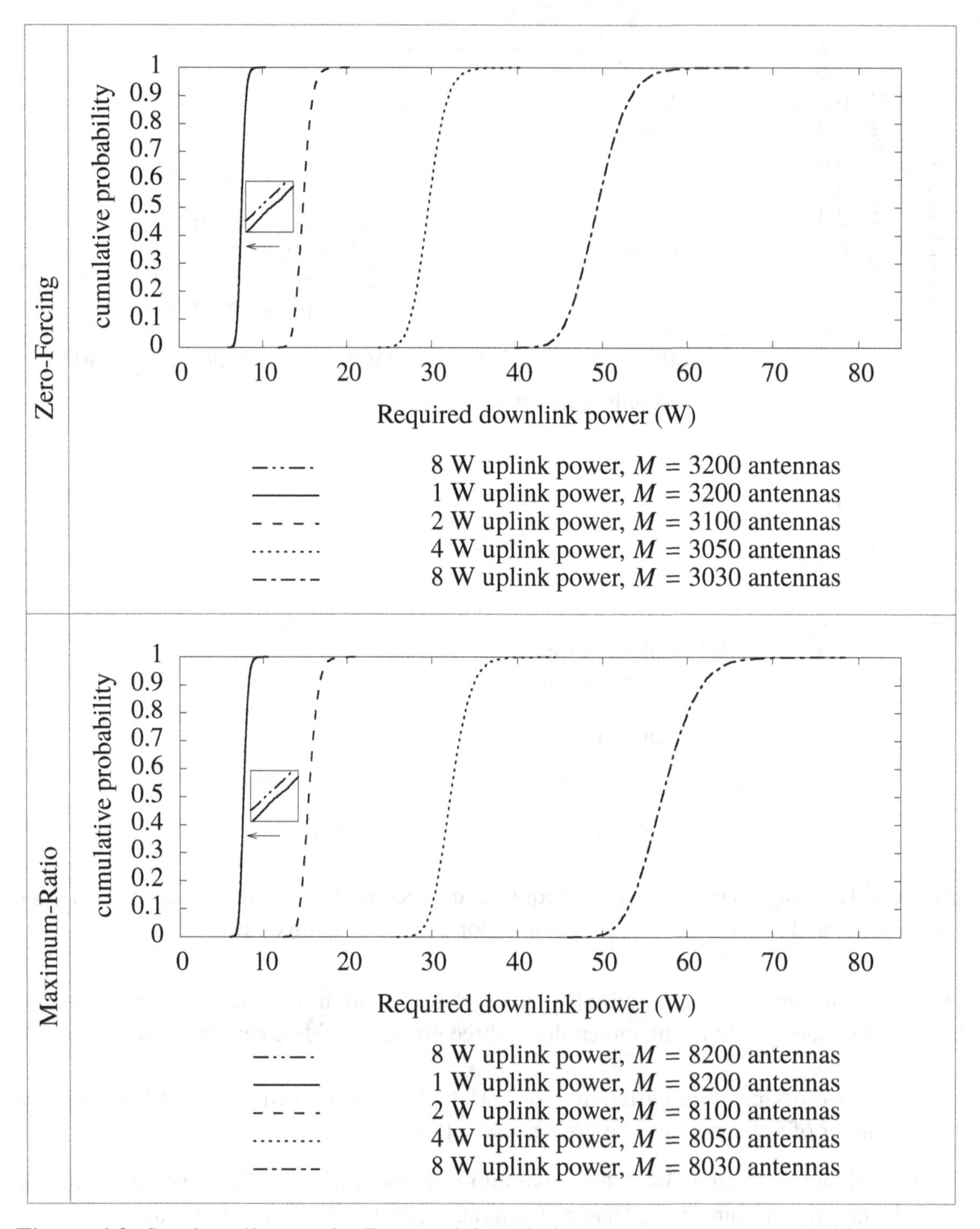

Figure 6.2. Single-cell example: Required downlink transmit power with zero-forcing and maximum-ratio processing, for a range of base station antennas, M, and uplink powers.

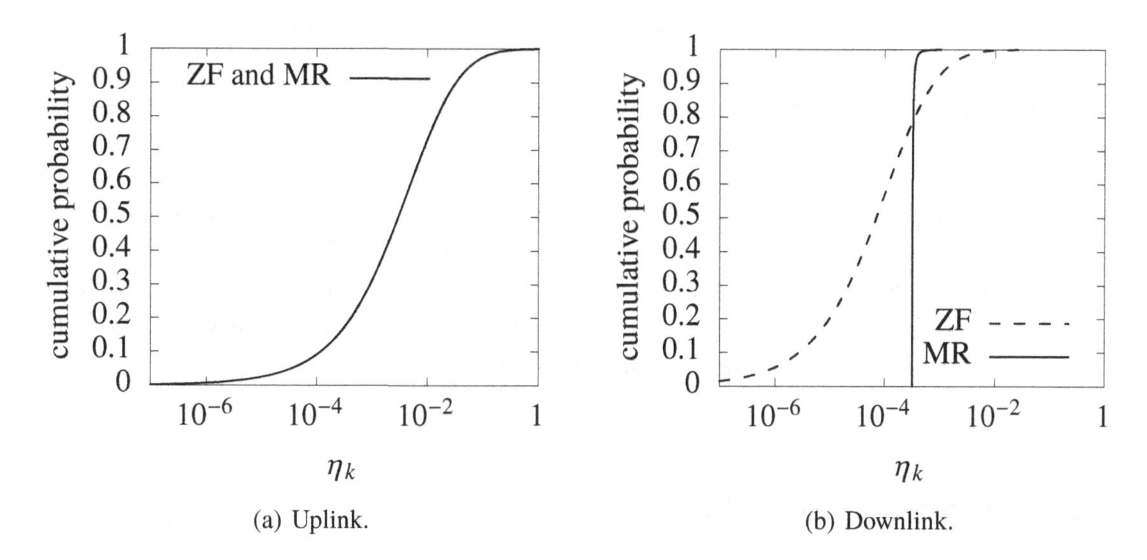

Figure 6.3. Single-cell example: Cumulative distribution of power control coefficients $\{\eta_k\}$ for zero-forcing (ZF) and maximum-ratio (MR) processing.

of the data. This yields a "squaring" effect such that reducing power by a factor of two requires the number of antennas to quadruple; see Section 3.4.3.

Figure 6.2 shows the cumulative distribution of the downlink power required to achieve a downlink rate of 20 Mb/s/terminal for uplink powers and numbers of antennas that are found sufficient to deliver the stipulated uplink throughput. The principal result to note is that for both zero-forcing and maximum-ratio processing, no more than 10 W of downlink power is required. For the reasons given above, the use of more power does not permit a significant reduction in the number of antennas.

Finally, to give some insight into the operating SNRs in this example, note that with an uplink radiated power of 1 W, we have $\rho_{\text{ul}} = 128$ dB (see calculation in Appendix F). At the cell edge, 11.3 km away from the base station, the Hata rural open model predicts a median path loss of $\beta = -125$ dB. Thus, the median uplink SNR for terminals at the cell edge is $\rho_{\text{ul}}\beta = 3$ dB. Similarly, with a downlink radiated power of 10 W, we have $\rho_{\text{dl}} = 138$ dB and a median downlink SNR at the cell edge of $\rho_{\text{dl}}\beta = 13$ dB.

6.1.2 Analysis of the Max-Min Fairness Power Control Policy

Figure 6.3 shows the distribution of the power control coefficients $\{\eta_k\}$ in the single-cell deployment example. With reference to Table 5.4, we observe the following:

- In the uplink, $\{\eta_k\}$ are identical for zero-forcing and maximum-ratio processing, and η_k is inversely proportional to γ_k. As a result, the spread of $\{\eta_k\}$ reflects the spread of $\{\gamma_k\}$. In turn, the long-duration pilots provide high-quality estimates for most of the terminals, so typically $\gamma_k \approx \beta_k$.

- In the downlink, under zero-forcing processing η_k is approximately inversely proportional to γ_k and the distribution of $\{\eta_k\}$ is qualitatively similar to that of uplink zero-forcing. For maximum-ratio processing, by contrast, since $\rho_{\mathrm{dl}}\beta_k \gg 1$ (13 dB median at the cell edge), η_k is approximately the same for all k. As a result, all terminals are allocated roughly the same power.

6.2 Multi-Cell Deployment: Preliminaries and Algorithms

A starting point for the design of multi-cell Massive MIMO systems is the theory presented in Chapters 4 and 5. As we shall see, a number of subsidiary issues have to be addressed, including multi-cell modeling, algorithms for assignment of terminals to base stations, and choice of appropriate uplink pilot reuse patterns.

6.2.1 Multi-Cell Cluster Modeling and Pilot Reuse

We assume a network of hexagonal cells, all of which fully reuse the same resources for data transmission. Pilot contamination emerges as a major problem which is most simply mitigated, with some additional overhead, by employing a pilot reuse factor, n_{reuse}, greater than one, so that all pilots in groups of n_{reuse} adjacent cells are orthogonal. The geometry of the pilot reuse strategy is the same as for traditional frequency division multiple access (FDMA) wireless networks [32–34], and common reuse factors are $n_{\mathrm{reuse}} = 1, 3, 4$ and 7.

Figures 6.4–6.6 illustrate the configurations that we employ for pilot reuse 7, 3, and 4 respectively. Since we want to observe residual pilot contamination, and the reuse factor should divide the total number of cells, reuse 7 requires a total of 49 cells, while reuse 3 or 4 requires 48 cells. All hexagons within a cluster tessellate, and the clusters themselves tessellate. This makes it possible to replicate the cluster periodically so that every cell in the cluster becomes statistically identical. That is, every cell is surrounded by other cells in the cluster in the same fashion, so that no cell is in a more advantageous position than any other. Clearly, for $n_{\mathrm{reuse}} = 1$, either the 49-cell or 48-cell cluster may be used, with substantially identical result. The cells are enumerated such that the nth cell and the $(n \pm n_{\mathrm{reuse}})$th cell use the same set of orthogonal pilots. The letters on the boundary edges indicate cell borders that coincide after periodic replication of the cluster.

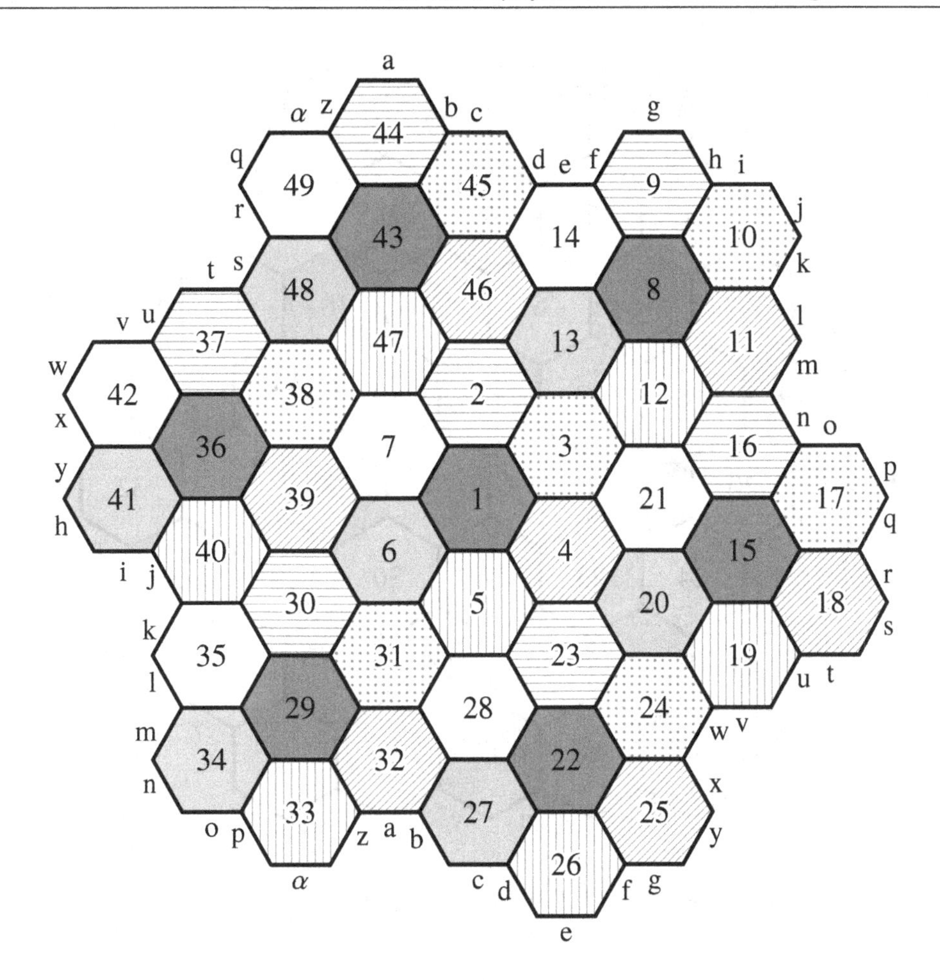

Figure 6.4. 49-cell cluster with pilot reuse $n_{\text{reuse}} = 7$. Each pattern is associated with a distinct set of orthogonal pilots that are also mutually orthogonal from pattern to pattern.

Reuse 7 surrounds the home cell with two concentric rings of non-contaminating cells, while reuse 3 and 4 surround the home cell with one ring of non-contaminating cells. The salutary reduction in pilot contamination comes with a price in terms of the proportionately longer pilot sequences that have to be used. If the number of active terminals, K, is small, this is not a problem, and large n_{reuse} can be afforded. For example, with a coherence interval of $\tau_c = 200$ samples, $K = 5$ simultaneously active terminals per cell, and reuse $n_{\text{reuse}} = 7$, the pilots will occupy at least $5 \times 7 = 35$ samples, which represents only a $35/200 = 17.5\%$ overhead. But with large K, the pilots may consume a substantial fraction of the available samples. For example, with $K = 30$ active terminals per cell and $n_{\text{reuse}} = 7$, $30 \times 7 = 210$ pilot symbols are needed, which is already more than the number of available samples (200).

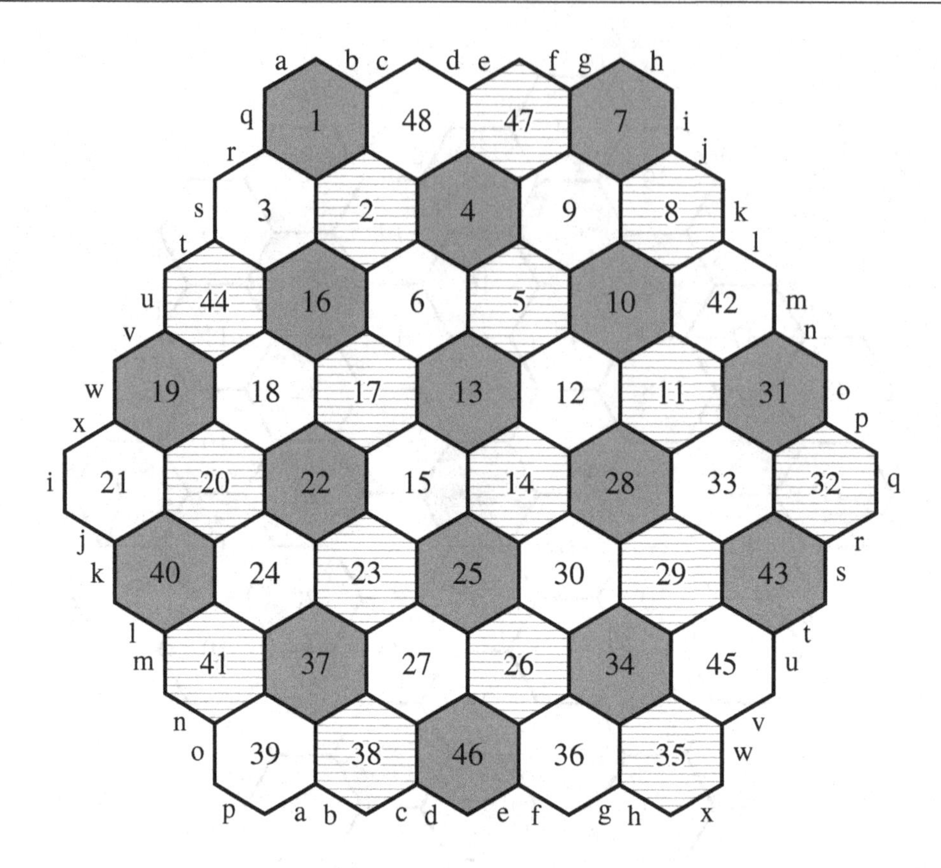

Figure 6.5. 48-cell cluster with pilot reuse $n_{\text{reuse}} = 3$.

6.2.2 Assignment of Terminals to Base Stations

In each cell, K terminals are randomly distributed within the hexagon centered at the base station. See Appendix G for an efficient algorithm to generate this distribution. Because of shadow fading, a given terminal may not be best served by the geographically closest base station, so the actual number of terminals served by a cell may deviate from K. In general, for a given terminal the best serving base station may be different for the uplink and the downlink. However, to avoid backhaul signaling overhead, it is desirable that the same base station jointly serves the uplink and the downlink for every terminal.

In all design examples, we assign each terminal to the base station to which it has the biggest large-scale fading coefficient. This strategy tends to favor the downlink over the uplink, because the uplink suffers greater variability of the non-coherent interference. In summary, our assignment strategy works well as a baseline, but better performance may be obtained with a more sophisticated algorithm.

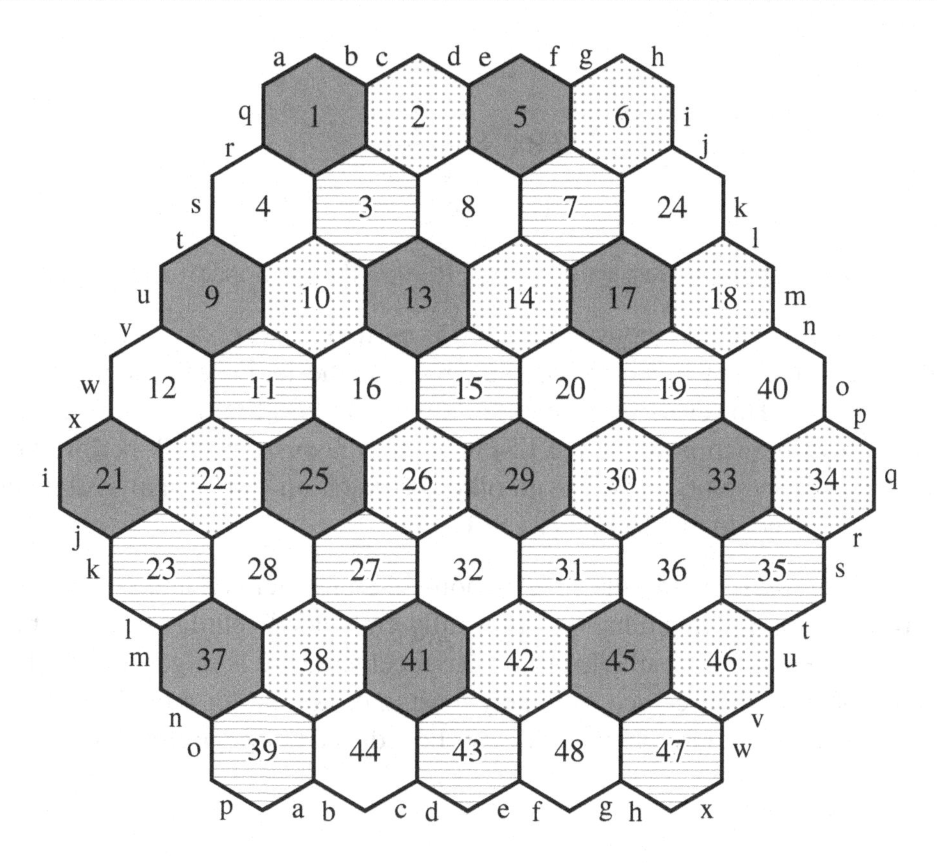

Figure 6.6. 48-cell cluster with pilot reuse $n_{\text{reuse}} = 4$.

6.2.3 Dropping Terminals from Service

In all wireless networks, a small percentage of the terminals will experience severe shadow fading and should be dropped from service, particularly if max-min power control is used. By dropping a small fraction of terminals from service, one can achieve substantial savings in power, or equivalently, a greatly reduced required number of base station antennas.

Typical coverage by premier wireless service providers in a macro-cellular deployment is 95%; that is, the 5% of the terminals with the worst channel conditions in terms of large-scale fading (path loss and shadow fading) are dropped from service. Ideally, the terminals with the lowest SINRs should be dropped. However, the SINR for each terminal, as shown in Chapter 5, can only be computed after the power control coefficients are determined, which in turn depend on which terminals are served. Thus, we need a simple means to estimate the channel conditions for each terminal before the power control coefficients are determined. For this purpose, we define the (downlink) *large-scale fading ratio* (LSFR) for each terminal as

$$\mathrm{LSFR}_{lk} = \frac{\beta_{lk}^{l}}{\sum\limits_{l'=1}^{L} \beta_{lk}^{l'}}. \tag{6.2}$$

The 5% terminals with the lowest value of LSFR_{lk} are then dropped from service.

The LSFR in (6.2) is a very convenient metric as it is independent of the uplink and downlink powers, the power control coefficients, and the large-scale fading coefficients of other terminals. However, other metrics are possible. Similar to the base station assignment policy in Section 6.2.2, the LSFR in (6.2) favors downlink performance over uplink performance to some degree as no other base-station-to-terminal assignment than that in Section 6.2.2 could yield a higher LSFR.

While the LSFR in (6.2) captures the effect of non-coherent inter-cell interference, it ignores receiver noise and pilot contamination. In a fully embedded cellular system where each base station radiates its full available power, the receiver noise is in general much smaller than the non-coherent interference from neighboring cells; and for a moderate number of antennas, pilot contamination can in many cases be reduced to a negligible level with proper pilot reuse.

6.2.4 Pilot Assignment and Required Pilot Sequence Length

If every cell served exactly K terminals, the minimum required duration of the pilot sequences, τ_{p}, would be equal to Kn_{reuse}. More generally, among the cells that reuse a particular set of pilots, some cells serve more terminals than others. In the $n_{\mathrm{reuse}} = 4$ configuration in Figure 6.6, for example, cells $\{3, 7, 11, \ldots, 47\}$ share the same pilots, and the number of orthogonal pilots required by this group of cells is equal to the maximum number of terminals that any of the cells is responsible for. For each of the n_{reuse} groups of pilots, $j = 1, \ldots, n_{\mathrm{reuse}}$, denote by $K_{\mathrm{max},j}$ the maximum number of terminals served by any of the cells in that group. Then the required pilot duration is equal to the total number of mutually orthogonal pilots in the system,

$$\tau_{\mathrm{p}} = \sum_{j=1}^{n_{\mathrm{reuse}}} K_{\mathrm{max},j}. \tag{6.3}$$

In each cell, every terminal is randomly assigned one of the pilot sequences allocated to that cell. Conceivably, an artful pilot assignment algorithm may perform better.

6.2.5 Per-Cell Max-Min Power Control that Accounts for Coherent Interference

In Section 5.3.3, we showed that by neglecting coherent interference (the fourth term in the denominator of (5.2)), and assuming that all cells use full power as made precise by (5.29) and (5.30), max-min SINR power control can be performed independently within each cell. In situations where coherent interference is significant, a better algorithm is needed. In what follows, we present a heuristic algorithm that corrects, to the first order of approximation, the power control coefficients specified in Section 5.3.3 for the effects of coherent interference.

We start with initial power control coefficients, $\{\hat{\eta}_{lk}\}$, obtained under the assumption that the coherent interference can be neglected as described in Section 5.3.3. Let $\widehat{\text{SINR}}_{lk}$ denote the resulting exact SINR according to (5.2). We note that the denominator in (5.2), which represents the effective noise, is the sum of LK terms, which suggests that perturbations in the power control coefficients should have little effect on the value of the denominator. Then, within the limits of this approximation, the quantity

$$\hat{f}_{lk} = \frac{\widehat{\text{SINR}}_{lk}}{\hat{\eta}_{lk}} \tag{6.4}$$

may be interpreted as the SINR that the kth terminal in the lth cell would obtain for $\eta_{lk} = 1$. Hence, for any set of power control coefficients, $\{\eta_{lk}\}$, the resulting SINR is

$$\begin{aligned}\text{SINR}_{lk} &\approx \left(\frac{\widehat{\text{SINR}}_{lk}}{\hat{\eta}_{lk}}\right)\eta_{lk} \\ &= \hat{f}_{lk}\eta_{lk}.\end{aligned} \tag{6.5}$$

Through this approximate formula, the (nearly) max-min optimal power control coefficients in each cell follow directly.

Uplink

Let $\{\hat{\eta}_{lk}\}$ be initial power control coefficients given by (5.32),

$$\hat{\eta}_{lk} = \frac{\min\limits_{k'}\{a_{lk'}\}}{a_{lk}}, \qquad k = 1,\ldots,K, \qquad l = 1,\ldots,L. \tag{6.6}$$

Then the use of the approximate formula (6.5) and the max-min principle yields the power control coefficients,

$$\eta_{lk} = \frac{\min\limits_{k'}\left\{\hat{f}_{lk'}\right\}}{\hat{f}_{lk}}. \tag{6.7}$$

The resulting SINR is

$$\begin{aligned}
\text{SINR}_{lk} &\approx \hat{f}_{lk}\eta_{lk} \\
&= \min_{k'} \left\{ \hat{f}_{lk'} \right\} \\
&= \min_{k'} \left\{ \frac{\widehat{\text{SINR}_{lk'}}}{\hat{\eta}_{lk'}} \right\}.
\end{aligned} \tag{6.8}$$

Downlink

When neglecting coherent interference, the power control coefficients (5.35) yield equal SINR within each cell,

$$\hat{\eta}_{lk} = \frac{1 + \sum_{l' \in \mathcal{P}_l} b_{lk}^{l'} + \sum_{l' \notin \mathcal{P}_l} c_{lk}^{l'}}{a_{lk} \sum_{k''=1}^{K} \frac{1 + \sum_{l' \in \mathcal{P}_l} b_{lk''}^{l'} + \sum_{l' \notin \mathcal{P}_l} c_{lk''}^{l'}}{a_{lk''}}}. \tag{6.9}$$

To account for pilot contamination, we adjust the initial coefficients $\{\hat{\eta}_{lk}\}$ obtained in (6.9) by setting

$$\eta_{lk} = \frac{\frac{1}{\hat{f}_{lk}}}{\sum_{k'=1}^{K} \frac{1}{\hat{f}_{lk'}}}. \tag{6.10}$$

The resulting SINR is

$$\begin{aligned}
\text{SINR}_{lk} &\approx \hat{f}_{lk}\eta_{lk} \\
&= \frac{1}{\sum_{k'=1}^{K} \frac{1}{\hat{f}_{lk'}}} \\
&= \frac{1}{\sum_{k'=1}^{K} \frac{\hat{\eta}_{lk'}}{\widehat{\text{SINR}_{lk'}}}}.
\end{aligned} \tag{6.11}$$

6.3 Multi-Cell Deployment Examples: Mobile Access

We give two examples of a multi-cell deployment of Massive MIMO: mobile access in a dense urban environment, and mobile access in a suburban environment. The requirement is that 95% of the terminals should be guaranteed the best max-min service possible, regardless of their location in the cell and subject to high mobility. We shall assume an average of $K = 18$ simultaneously active terminals in each hexagonal cell.

In both examples, the system operates at a carrier frequency of 1.9 GHz and uses a total bandwidth of 20 MHz. After excluding the samples used for pilots, half of the coherence interval is used for the uplink and half for the downlink.

The base station array is located 30 meters above the ground and radiates 1 W in the downlink. The radiated power by each terminal in the uplink is 200 mW. The pilots are always transmitted at this maximum power level, while the uplink data transmission is subject to power control.

6.3.1 Dense Urban Scenario

In the dense urban deployment, a base station array of $M = 64$ antennas serves a cell of radius (center to vertex) 500 m. Such arrays can have small form factors [35]. The Hata-COST231 propagation model for metropolitan areas [36] predicts a median path loss of 129 dB at the cell edge.[1]

We assume a coherence time of $T_c = 2$ ms, which theoretically permits a mobility of 142 km/h, that is, 71 km/h with a factor-of-two design margin; see Section 2.1.[2] With a coherence bandwidth of $B_c = 210$ kHz, the coherence interval is of length $\tau_c = B_c T_c = 420$. The system employs OFDM, with the parameters shown in Table 2.2; hence there are $(14/15) \times 420 = 392$ useful samples per coherence interval when the cyclic prefix is accounted for. Instantaneous spectral efficiency is converted to net spectral efficiency, on either the uplink or the downlink, by applying the conversion factor $(1/2) \times (14/15) \times (1 - \tau_p/392)$, where the factor 1/2 represents the even split of data transmission time between uplink and downlink.

6.3.2 Suburban Scenario

In the suburban deployment, the base station has $M = 256$ antennas and serves a cell of radius 2 km. At 1.9 GHz carrier frequency, the Hata-COST231 propagation model for medium-sized cities [36] predicts 135 dB path loss between the base station antenna array

[1]The nominal applicable range for the Hata-COST231 model is from 1 km to 20 km. We have, however, used the same formula for ranges less than 1 km, as is usual in RF prediction tools.

[2]Numerical results in some earlier research papers by the authors used a different convention for the definition of coherence time.

and the cell edge.[3] This path loss is 6 dB larger than in the dense urban case, and we compensate for that by increasing M from 64 to 256.

A mobility of up to 284 km/h is allowed for, that is, 142 km/h with a factor-of-two design margin, see Section 2.1.4, resulting in a coherence time of $T_c = 1$ ms. With a coherence bandwidth of $B_c = 210$ kHz, the length of the coherence interval is $\tau_c = B_c T_c = 210$ samples, of which 196 are useful. All instantaneous spectral efficiencies are multiplied by $(1/2) \times (14/15) \times (1 - \tau_p/196)$ to produce net spectral efficiencies.

6.3.3 Minimum Per-Terminal Throughput Performance

The choice of pilot reuse factor entails a tradeoff between a reduction in the deleterious effects of pilot contamination, and additional time expended on transmission of pilots. In what follows, we investigate the impact of pilot reuse factor $n_{\text{reuse}} = 1, 3, 4, 7$ in the configurations of cells shown in Figures 6.4–6.6 under the power control algorithm of Section 6.2.5. Recall that this strategy employs maximum power in each cell, and it approximately achieves max-min SINR power control in each cell. Throughput is nearly the same for all terminals in a given cell, but throughput can differ considerably from cell to cell.

Figures 6.7 and 6.8 show cumulative distributions of the minimum per-terminal throughput in each cell derived from 100 independent random realizations of terminal positions and shadow fading. For each of the 100 realizations, the smallest throughput in each of the 48 or 49 cells is identified for generation of the cumulative distribution. The principal conclusion is that pilot reuse 7 is best for dense urban, while reuse 3 – marginally better than reuse 4 – is best for suburban. This conclusion holds not only with respect to median throughput (50th percentile), but also for the arguably more important 95% likely throughput (5th percentile). In no case is reuse 1 a good idea: optimum reuse increases the 95% likely throughput by a factor of at least 1.6, and in the case of dense urban uplink with maximum-ratio processing by a factor of 4.8. A significant finding is that in all cases non-coherent interference is the main impairment, and zero-forcing and maximum-ratio processing yield comparable performance.

Note that the 95% likely throughput is higher in the downlink than in the uplink, despite the fact that the total available power in the uplink (18×0.2 W per cell on average) exceeds that of the downlink (1 W per cell). The reason is that with the power control policy in Section 6.2.5, the most disadvantaged terminal in each cell transmits with full power and this terminal determines the common uplink SINR in the cell. Some of the terminals in the cell use very little uplink power. On the downlink, the base station can, in effect, borrow

[3] See Footnote 1.

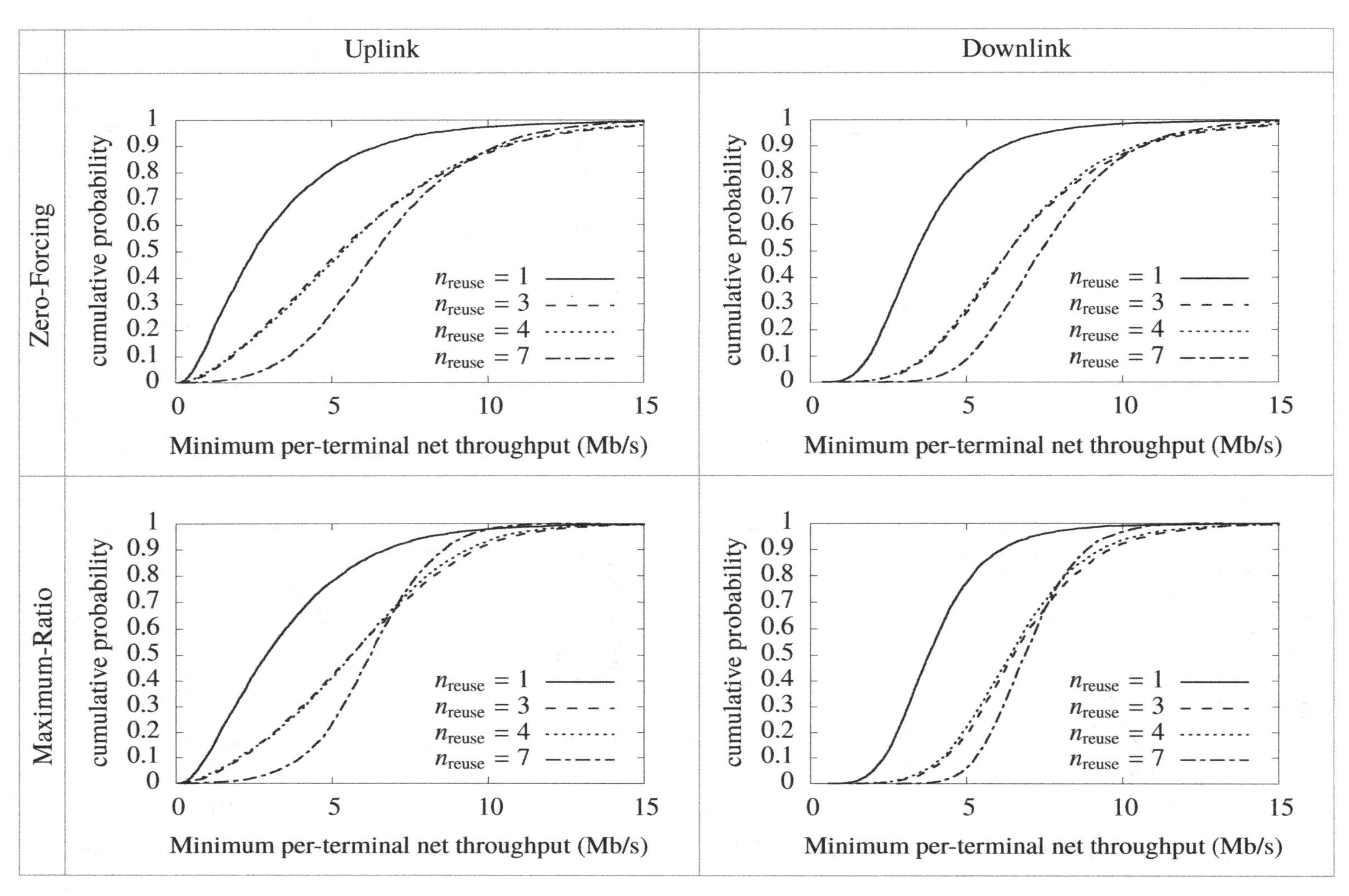

Figure 6.7. Dense urban deployment: Minimum per-terminal net throughput with zero-forcing and maximum-ratio processing, for different pilot reuse factors n_{reuse}, using the power control algorithm in Section 6.2.5.

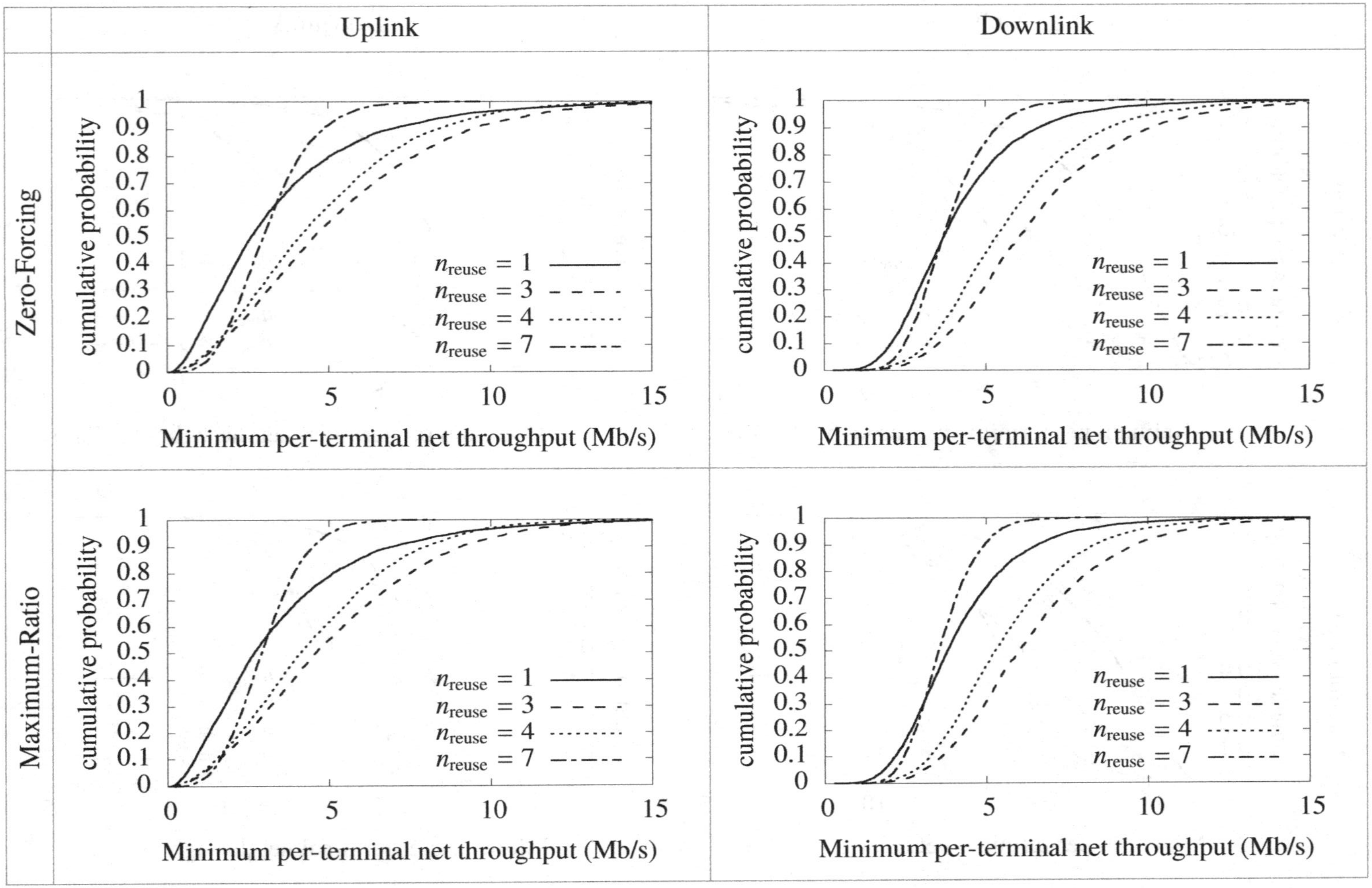

Figure 6.8. Suburban deployment: Minimum per-terminal net throughput with zero-forcing and maximum-ratio processing, for different pilot reuse factors n_{reuse}, using the power control algorithm in Section 6.2.5.

power from one terminal and give it to another, an option not available on the uplink.

6.3.4 Additional Observations

From Figures 6.7 and 6.8, and from additional numerical experiments not shown here, we have made the following additional observations:

- Increasing the uplink pilot power beyond 200 mW has a limited effect on the per-terminal throughput. Hence, the received SNR in the pilots is relatively high and the main impairment to channel estimation quality is pilot interference from other cells.

  The non-coherent inter-cell interference – given by the second and third terms in the denominator of (5.2) – is the dominant impairment on average, and it significantly exceeds the coherent interference power, the fourth term in the denominator of (5.2). However, the impact of pilot contamination on the strength of the useful signal, the numerator of (5.2), is significant. This is so because a_{lk} in (5.2) is proportional to γ_{lk}^{l} in (4.4). In the absence of pilot contamination, and with a high uplink SNR, γ_{lk}^{l} is close to β_{lk}^{l}. However, with a small pilot reuse factor, pilot contamination can cause γ_{lk}^{l} to be much smaller than β_{lk}^{l}. Increasing the pilot reuse factor, n_{reuse}, reduces $\sum_{l'' \in \mathcal{P}_l} \beta_{l''k}^{l}$ in the denominator in (4.4) and hence brings γ_{lk}^{l} closer to β_{lk}^{l}.

  In summary, increasing $\tau_{\text{p}}\rho_{\text{ul}}$ has limited impact because it increases both the numerator and the denominator in (4.4) by roughly the same amount, whereas increasing n_{reuse} has a significant impact because it reduces the denominator in (4.4) while keeping the numerator unchanged.

- In the downlink, increasing the radiated power from 1 W to 10 W can result in a 95% likely throughput performance increase by about 50%. This means that with 1 W output power, the system is not yet entirely interference limited in the downlink. This is so because the cell radii in our case study are relatively large for the 1.9 GHz band. We limited the downlink radiated power to 1 W to show that Massive MIMO can offer superior radiated energy efficiency, and to obtain comparable downlink and uplink throughput performance with an equal uplink/downlink split of the coherence interval. In practice, lower radiated power may simplify the design of certain hardware components, such as channel filters.

  Similarly, increasing the uplink radiated power beyond 200 mW can also result in a noticeable throughput performance gain.

6.3.5 Comparison of Power Control Policies

The performance examples of Figures 6.7 and 6.8 used the heuristic power control algorithm of Section 6.2.5, with full downlink power in each cell, full uplink power by at least one terminal, and which approximately equalizes throughput within each cell. We now compare the performance of this power control with that of five other algorithms, described below (note that the term used in the figure legend is underlined):

(i) *Equal-η*: Equal and maximum permitted power for all terminals. In the uplink, $\eta_{lk} = 1$ for all l and k. In the downlink, $\eta_{lk} = 1/K$ for all l and k.

(ii) *Equal-SNR*: Coherent beamforming gain (useful received power) is equalized within every cell. In the uplink,

$$\eta_{lk} = \frac{\min_{k'} \left\{ \gamma^l_{lk'} \right\}}{\gamma^l_{lk}} \tag{6.12}$$

for all l and k and in the downlink,

$$\eta_{lk} = \frac{1}{\gamma^l_{lk} \sum_{k'=1}^{K} \frac{1}{\gamma^l_{lk'}}} \tag{6.13}$$

for all l and k. There is no particular motivation behind this policy, other than that it is simple and potentially tempting to use.

(iii) *Network-wide equal-SINR*: The network-wide max-min fairness power control algorithm of Section 5.3.2. As pointed out, it is not scalable as the number of cells in the network increases. However, it is still useful as a baseline for comparisons.

(iv) *Single-cell equal-SINR*: The max-min fairness power control algorithm for a single-cell scenario, given in Section 5.3.1 and summarized in Table 5.4, neglects all interference from cells other than the home cell.

(v) *Multi-cell intra-cell equal-SINR, neglecting coherent interference (CI)*: The algorithm of Section 5.3.3, summarized in Tables 5.5 and 5.6, seeks max-min SINR fairness within each cell while neglecting coherent interference and operating all cells at full power.

In the uplink, this policy is equivalent to the equal-SNR power control in (ii) above. This follows by comparing the expression for $\{\eta_{lk}\}$ in Table 5.5 to (6.12). Hence, for the uplink, curves for this power control policy are omitted in the plots.

(vi) *Multi-cell approximate intra-cell equal-SINR, with correction for coherent interference*: The algorithm in Section 6.2.5, also used in the comparisons in Figures 6.7 and 6.8.

Figures 6.9 and 6.10 show the performance of the six power control algorithms for dense urban ($n_{\text{reuse}} = 7$) and suburban ($n_{\text{reuse}} = 3$) deployments respectively. We observe the following:

- Power control is exceedingly important for Massive MIMO. There is a huge gap between the better-performing power controls and *equal-η*, which is substantially no power control at all.
- The *multi-cell approximate intra-cell equal-SINR with correction for coherent interference* policy consistently outperforms all other power control schemes examined here. In particular, it yields improvements over the algorithms that neglect coherent interference, especially in the suburban environment where n_{reuse} is smaller.
- Power control that accounts for coherent interference has greater impact on the downlink than on the uplink, because ρ_{dl} is 7 dB bigger than ρ_{ul}, and the coherent interference in Table 4.1 scales with these parameters.
- *Equal-SNR* power control performs rather well on uplink, and comparatively poorly on downlink, because on uplink the non-coherent interference is the same for all terminals in the same cell. On the downlink, zero-forcing performs somewhat better than maximum-ratio because of reduced intra-cell interference.
- In the uplink, *equal-SNR* power control performs slightly better than *single-cell equal-SINR*. The distinction between them is that the *equal-SNR* policy uses γ_{lk}^{l} in (4.4), whereas *single-cell equal-SINR* uses γ_k given by (3.8).

  Also in the uplink, the *multi-cell approximate intra-cell equal-SINR with correction for coherent interference* policy is effectively a modified version of the *equal-SNR* scheme, obtained by taking into account coherent interference. When the magnitude of the coherent interference is small (e.g., with $n_{\text{reuse}} = 7$ in the dense urban scenario), then *equal-SNR* is almost as good as *multi-cell approximate intra-cell equal-SINR with correction for coherent interference.* However, when coherent interference is significant, the performances of these two power control schemes differ appreciably.

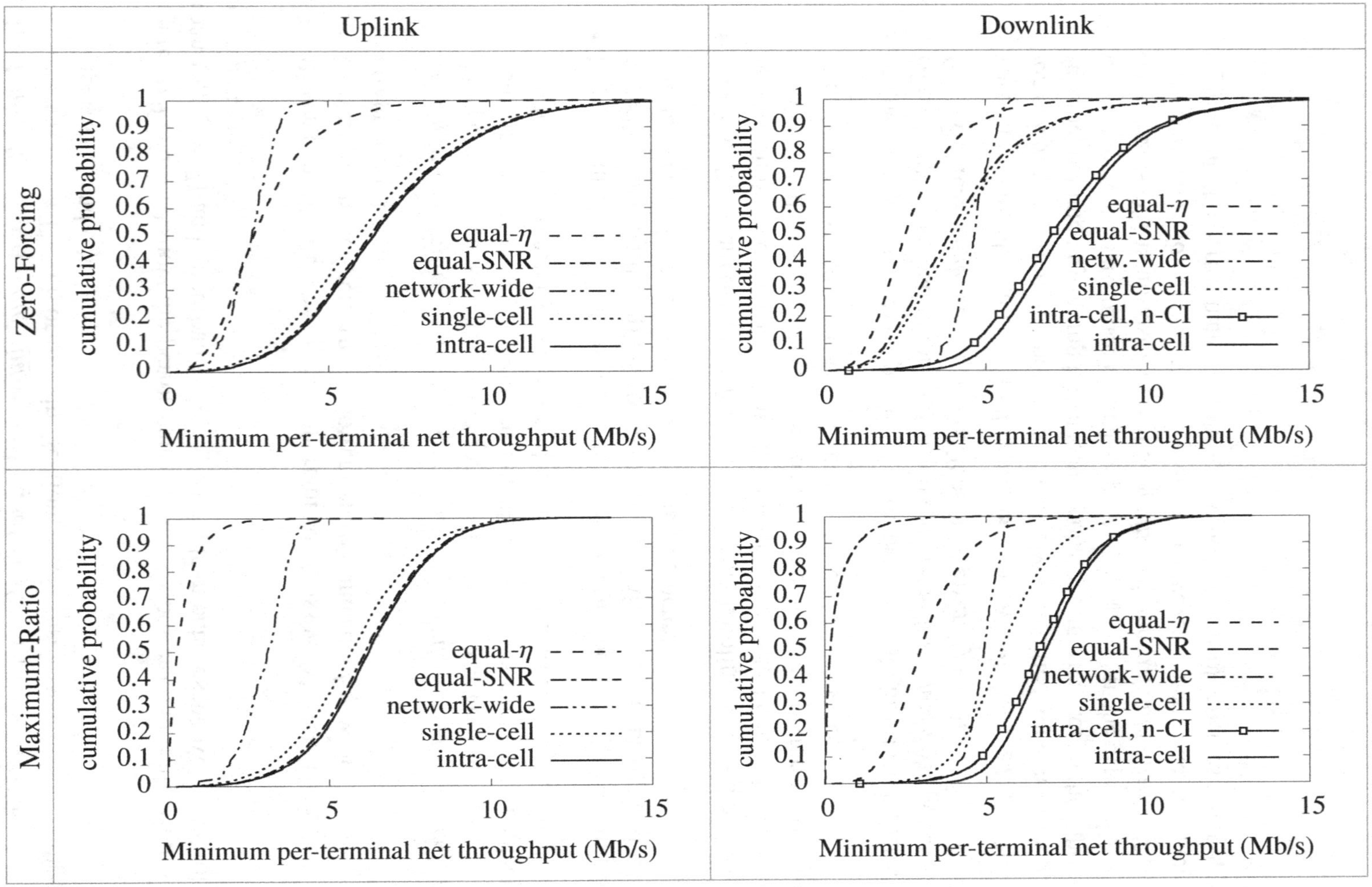

Figure 6.9. Dense urban deployment: Comparison of minimum per-terminal net throughput for six power control algorithms with zero-forcing and maximum-ratio processing, and $n_{\text{reuse}} = 7$.

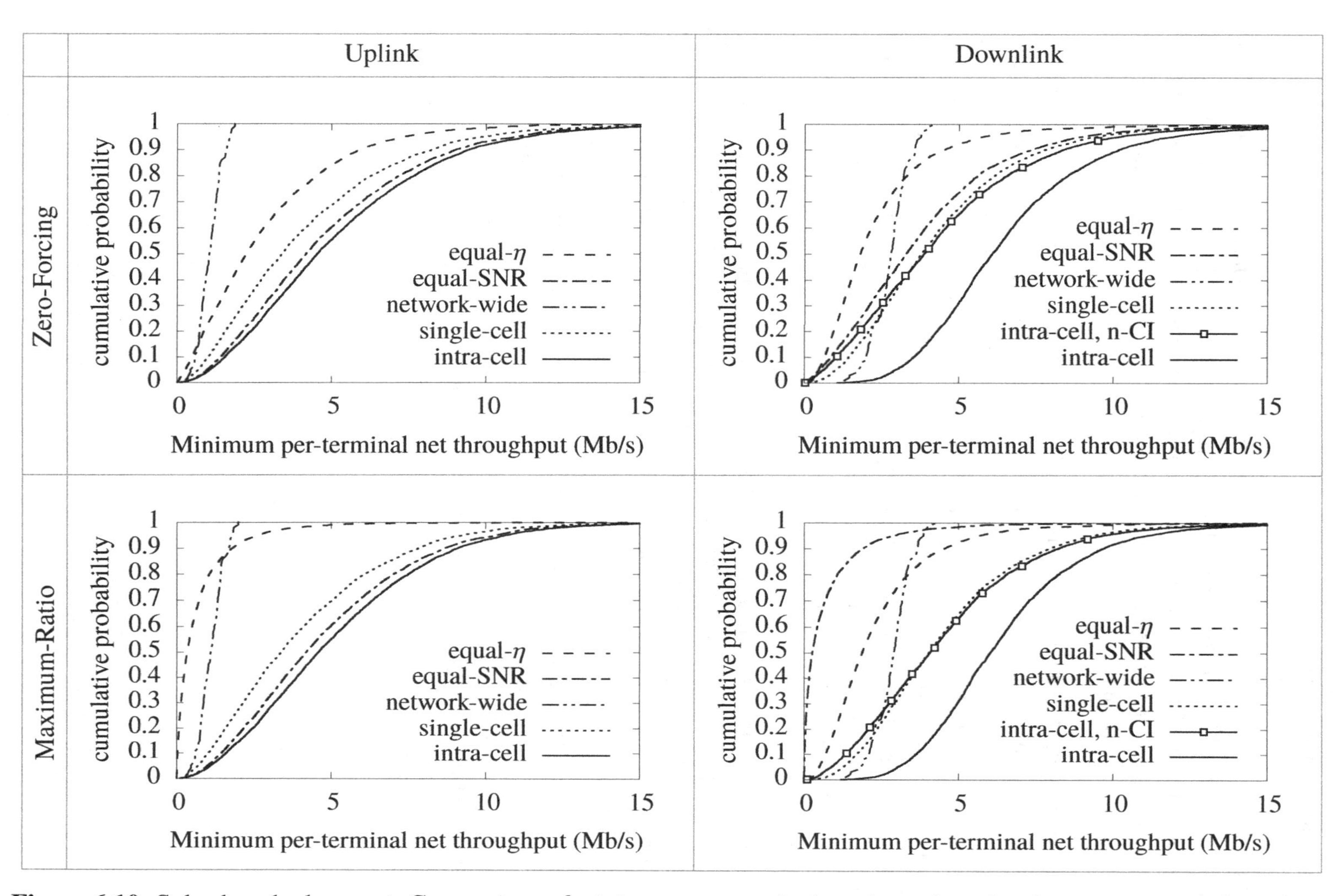

Figure 6.10. Suburban deployment: Comparison of minimum per-terminal net throughput for six power control algorithms with zero-forcing and maximum-ratio processing, and $n_{\text{reuse}} = 3$.

6.4 Summary of Key Points

- This chapter has showcased prime application examples of Massive MIMO technology: cable/fiber-like throughput to thousands of homes in a large geographical area, and mobile cellular service for both dense urban and suburban deployments. Table 6.1 details the parameter values for these scenarios, and Tables 6.2 and 6.3 summarize the resulting performance.

- When deploying Massive MIMO in multi-cell systems, pilots must be reused to accommodate many simultaneously active terminals. A higher pilot reuse factor, n_{reuse}, pushes cells that use the same pilots farther apart, resulting in lower pilot contamination and a higher effective SINR. But reusing pilots more sparsely means a higher overhead in terms of the number of samples in each coherence interval that are spent on pilots. Different choices of n_{reuse} can give very different performance, and hence n_{reuse} must be carefully chosen for each given deployment scenario. In particular, there is an optimal value of n_{reuse} that maximizes the per-terminal throughput.

  Our models suggest that in a dense urban environment where the cells are small and the terminal mobility is moderate, $n_{\text{reuse}} = 7$ is best; in a suburban deployment, where the cells are larger and terminal mobility is higher, $n_{\text{reuse}} = 3$ is best. In both environments, $n_{\text{reuse}} = 1$ delivers substantially lower throughputs and thus should be avoided.

- Careful power control is vital. Despite the simple dependence of SINR on power control coefficients, the optimization of power control for a multi-cell system is surprisingly difficult. The approximate algorithm described in Section 6.2.5 shows promise, but further improvements may be possible.

Chapter 7

THE MASSIVE MIMO PROPAGATION CHANNEL

All of our performance analyses are based on the i.i.d. Rayleigh model for fading under which the small-scale fading coefficients between every base station antenna and every terminal are i.i.d. $\mathcal{CN}(0, 1)$ random variables. The justification is, first, that this model is approximately correct under conditions of dense scattering, and second, that the model enables the comprehensive performance analyses of Chapters 3 and 4. In this chapter, we revisit small-scale fading, and we establish, under a radically different model, that Massive MIMO still functions well.

7.1 Favorable Propagation and Deterministic Channels

We begin by considering Massive MIMO operation in a single cell, as in Section 2.2.1, and we address two fundamental questions. For given constraints on the channel norms $\{\|\boldsymbol{g}_k\|\}$, we identify the joint behavior of the channel vectors that maximizes performance from, first, an information-theoretic perspective, and, second, from the perspective of linear processing.

Intuitively, to maximize performance, the channel vectors $\{\boldsymbol{g}_k\}$ should be as different as possible, according to some appropriate metric. To make this notion precise, we say that the channel offers *favorable propagation* if

$$\boldsymbol{g}_k^{\mathrm{H}} \boldsymbol{g}_{k'} = 0, \qquad k, k' = 1, \ldots, K, \qquad k' \neq k. \tag{7.1}$$

In the following sections, we shall see why (7.1) represents the most favorable scenario. In practice, (7.1) will never be exactly satisfied, but it can be *approximately* satisfied and we then say that we have *approximately favorable propagation*. Also, under some assumptions on the propagation environment it holds asymptotically that

$$\frac{1}{M} \boldsymbol{g}_k^{\mathrm{H}} \boldsymbol{g}_{k'} \to 0, \quad M \to \infty, \quad k, k' = 1, \ldots, K, \quad k' \neq k. \tag{7.2}$$

When (7.2) is satisfied, we say that the environment offers *asymptotically favorable propagation.* Of course, the limiting case that $M \to \infty$ has no physical meaning, but in many cases, taking the limit is useful in order to understand the behavior of the propagation when M is large.

7.1.1 Capacity Upper Bound

The mutual orthogonality in (7.1) of $\{\boldsymbol{g}_k\}$ offered in favorable propagation represents the most desirable scenario from the perspective of maximizing the rate. To establish this fact, we use a capacity argument. Consider the uplink. From Section 1.2, we know that for a fixed, deterministic channel matrix $\boldsymbol{G}$, the sum capacity is

$$C_{\text{sum}} = \log_2 \left| \boldsymbol{I}_M + \rho_{\text{ul}} \boldsymbol{G}\boldsymbol{G}^{\mathrm{H}} \right|, \tag{7.3}$$

assuming that the base station knows $\boldsymbol{G}$ and that the terminals know their respective individual rates.

We now determine the largest value that C_{sum} in (7.3) can assume for specified channel norms $\{\|\boldsymbol{g}_k\|^2\}$. Analogously to (3.69), by using Sylvester's determinant theorem and the Hadamard inequality,

$$\begin{aligned} C_{\text{sum}} &= \log_2 \left| \boldsymbol{I}_M + \rho_{\text{ul}} \boldsymbol{G}\boldsymbol{G}^{\mathrm{H}} \right| \\ &= \log_2 \left| \boldsymbol{I}_K + \rho_{\text{ul}} \boldsymbol{G}^{\mathrm{H}}\boldsymbol{G} \right| \\ &\overset{(a)}{\leq} \log_2 \left(\prod_{k=1}^{K} [\boldsymbol{I}_K + \rho_{\text{ul}} \boldsymbol{G}^{\mathrm{H}}\boldsymbol{G}]_{kk} \right) \\ &= \sum_{k=1}^{K} \log_2 \left(1 + \rho_{\text{ul}} \|\boldsymbol{g}_k\|^2\right), \end{aligned} \tag{7.4}$$

with equality in (a) if and only if $\boldsymbol{G}^{\mathrm{H}}\boldsymbol{G}$ is diagonal, which is equivalent to (7.1). Hence, for given $\{\|\boldsymbol{g}_k\|^2\}$, C_{sum} is maximized when (7.1) holds. This confirms the soundness of favorable propagation as defined by (7.1).

The concept of favorable propagation can also be analyzed for the downlink, but this is considerably more difficult because the corresponding capacity expression involves solving an optimization problem.

7.1.2 Distance from Favorable Propagation

An important question is how far from favorable propagation a given channel matrix $\boldsymbol{G}$ is. One way to quantify the departure from favorable propagation is the ratio between C_{sum} and the upper bound in (7.4),

$$\Delta_{\text{C}} = \frac{\log_2 \left|\boldsymbol{I}_M + \rho_{\text{ul}}\boldsymbol{G}\boldsymbol{G}^{\text{H}}\right|}{\sum_{k=1}^{K} \log_2 \left(1 + \rho_{\text{ul}}\|\boldsymbol{g}_k\|^2\right)}. \tag{7.5}$$

In favorable propagation, $\Delta_{\text{C}} = 1$. An operationally equivalent measure is the power increase, say $\Delta_{\rho_{\text{ul}}}$ that would be needed for the sum capacity offered by $\boldsymbol{G}$ to reach the upper bound in (7.4) – that is, the value of $\Delta_{\rho_{\text{ul}}}$ that solves the following equation:

$$\sum_{k=1}^{K} \log_2 \left(1 + \rho_{\text{ul}}\|\boldsymbol{g}_k\|^2\right) = \log_2 \left|\boldsymbol{I}_M + \Delta_{\rho_{\text{ul}}}\rho_{\text{ul}}\boldsymbol{G}\boldsymbol{G}^{\text{H}}\right|. \tag{7.6}$$

Note that both Δ_{C} and $\Delta_{\rho_{\text{ul}}}$ depend on the SNR, ρ_{ul}.

7.1.3 Favorable Propagation and Linear Processing

We have seen in Chapters 3 and 4 that zero-forcing and maximum-ratio processing are exceedingly effective. Their performance is bounded by that of the minimum mean-square error (MMSE) filter. In turn, we can establish the propagation conditions that most enhance MMSE performance.

Consider the uplink, given by (2.27), with the assumption that $\{x_k\}$ are i.i.d. CN(0, 1). The objective of the base station is to detect $\boldsymbol{x}$, given $\boldsymbol{y}$. The MMSE estimate of $\boldsymbol{x}$ is

$$\begin{aligned}\hat{\boldsymbol{x}}_{\text{mmse}} &= \mathsf{E}\{\boldsymbol{x}|\boldsymbol{y}\} \\ &= \sqrt{\rho_{\text{ul}}}\boldsymbol{G}^{\text{H}}\left(\boldsymbol{I}_M + \rho_{\text{ul}}\boldsymbol{G}\boldsymbol{G}^{\text{H}}\right)^{-1}\boldsymbol{y} \\ &= \sqrt{\rho_{\text{ul}}}\left(\boldsymbol{I}_K + \rho_{\text{ul}}\boldsymbol{G}^{\text{H}}\boldsymbol{G}\right)^{-1}\boldsymbol{G}^{\text{H}}\boldsymbol{y}.\end{aligned} \tag{7.7}$$

The error covariance of $\hat{\boldsymbol{x}}_{\text{mmse}}$ is

$$\begin{aligned}\mathsf{Cov}\{\boldsymbol{x}|\boldsymbol{y}\} &= \boldsymbol{I}_K - \rho_{\text{ul}}\boldsymbol{G}^{\text{H}}\left(\boldsymbol{I}_M + \rho_{\text{ul}}\boldsymbol{G}\boldsymbol{G}^{\text{H}}\right)^{-1}\boldsymbol{G} \\ &= \left(\boldsymbol{I}_K + \rho_{\text{ul}}\boldsymbol{G}^{\text{H}}\boldsymbol{G}\right)^{-1}.\end{aligned} \tag{7.8}$$

The $\{x_k\}$ are statistically independent, so the transmissions by the other terminals can only hurt the detection of x_k. Hence an upper bound on the performance of any detection scheme corresponds to the hypothetical scenario whereby the kth terminal alone transmits, and maximum-ratio processing is optimal,

$$\hat{x}_{\text{mmse},k}\Big|_{k\text{th terminal alone}} = \frac{\sqrt{\rho_{\text{ul}}}}{1+\rho_{\text{ul}}\|\boldsymbol{g}_k\|^2}\boldsymbol{g}_k^{\text{H}}\boldsymbol{y}. \tag{7.9}$$

This establishes an upper bound on performance,

$$\mathsf{Var}\{x_k|\boldsymbol{y}\} \geq \frac{1}{1+\rho_{\text{ul}}\|\boldsymbol{g}_k\|^2}. \tag{7.10}$$

Finally, we note that if the channel vectors were perfectly orthogonal, see (7.1), then maximum-ratio processing would attain the above performance bound. Put another way, mutually orthogonal channel vectors represent the best possible propagation, in which case maximum-ratio processing (and incidentally zero-forcing) is optimal.

7.1.4 Singular Value Spread as a Measure of Favorable Propagation

To gain some additional insight, note that many of the preceding equations can be expressed in terms of the singular values $\{\sigma_k\}$ of $\boldsymbol{G}$. Particularly, for the sum capacity,

$$\begin{aligned} C_{\text{sum}} &= \log_2\left|\boldsymbol{I}_M + \rho_{\text{ul}}\boldsymbol{G}\boldsymbol{G}^{\text{H}}\right| \\ &= \log_2\left|\boldsymbol{I}_K + \rho_{\text{ul}}\boldsymbol{G}^{\text{H}}\boldsymbol{G}\right| \\ &= \sum_{k=1}^{K}\log_2\left(1+\rho_{\text{ul}}\sigma_k^2\right). \end{aligned} \tag{7.11}$$

Also, for the MMSE detector in (7.7),

$$\begin{aligned} \sum_{k=1}^{K}\mathsf{Var}\{x_k|\boldsymbol{y}\} &= \mathsf{Tr}\left\{\left(\boldsymbol{I}_K + \rho_{\text{ul}}\boldsymbol{G}^{\text{H}}\boldsymbol{G}\right)^{-1}\right\} \\ &= \sum_{k=1}^{K}\frac{1}{1+\rho_{\text{ul}}\sigma_k^2}. \end{aligned} \tag{7.12}$$

Note that there is no direct correspondence between the K singular values and the K terminals, except for the special case of favorable propagation where $\boldsymbol{G}^{\text{H}}\boldsymbol{G}$ is a diagonal matrix and $\sigma_k = \|\boldsymbol{g}_k\|$.

Let $\sigma_{\max}$ and $\sigma_{\min}$ be the extreme values of $\{\sigma_k\}$. If $\sigma_{\max} = \sigma_{\min}$, then we must have favorable propagation, because then $\boldsymbol{G}^{\mathrm{H}}\boldsymbol{G}$ is a scaled identity matrix. It is also clear that if we have asymptotically favorable propagation and if all $\{\beta_k\}$ are equal (to β_1, say), then

$$\frac{\sigma_{\max}^2}{M} \to \beta_1, \qquad \text{and} \qquad \frac{\sigma_{\min}^2}{M} \to \beta_1, \qquad M \to \infty, \tag{7.13}$$

so $\sigma_{\max}/\sigma_{\min} \approx 1$ for large M. Hence, in case all $\{\beta_k\}$ are equal, the singular value spread of $\boldsymbol{G}$ can be viewed as a proxy for how favorable $\boldsymbol{G}$ is.

The singular value ratio $\sigma_{\max}/\sigma_{\min}$ is simple to evaluate, and it does not depend on ρ_{ul}. However, the measure $\sigma_{\max}/\sigma_{\min}$ only has a meaning if $M \gg 1$ and all $\{\beta_k\}$ are equal, and it disregards all other singular values than $\sigma_{\min}$ and $\sigma_{\max}$. Hence, it is normally preferable to directly work with Δ_{C} in (7.5) or $\Delta_{\rho_{\mathrm{ul}}}$ in (7.6).

7.2 Favorable Propagation and Random Channels

So far in this chapter, we have considered favorable propagation for a given, deterministic channel matrix $\boldsymbol{G}$. In practice, $\boldsymbol{G}$ will be random due to fading. It is then of interest to examine to what extent we have favorable propagation "on average". To that end, we may look at the distribution of $\{\sigma_k\}$, or the probability that $\sigma_{\max}/\sigma_{\min}$ falls below a given threshold, or the probability that Δ_{C} and $\Delta_{\rho_{\mathrm{ul}}}$ fall below given thresholds. Alternatively, we may look at the inner products $\boldsymbol{g}_k^{\mathrm{H}}\boldsymbol{g}_{k'}$ in (7.1) on average.

Many, entirely different, practical scenarios result in approximately favorable propagation. To understand this, we will consider two particular cases that represent disparate physical situations: independent Rayleigh fading and uniformly random line-of-sight (UR-LoS). Throughout, we consider a single cell and assume that the base station array is uniform and linear with an antenna spacing of $\lambda/2$.

7.2.1 Independent Rayleigh Fading

The first scenario of interest is where the system operates in a dense, isotropic scattering environment; see Figure 7.1. We model this scenario by assuming that $\boldsymbol{G}$ has independent random elements $\{g_k^m\}$ with zero mean and distribution $\mathrm{CN}(0, \beta_k)$, and we refer to the scenario as *independent Rayleigh fading*.

The Gaussian distribution of $\{g_k^m\}$ can be justified by the central limit theorem, assuming that each antenna sees the superposition of many wavefronts that originate from independent scatterers. Completely independent Rayleigh fading is, strictly speaking, incompatible with

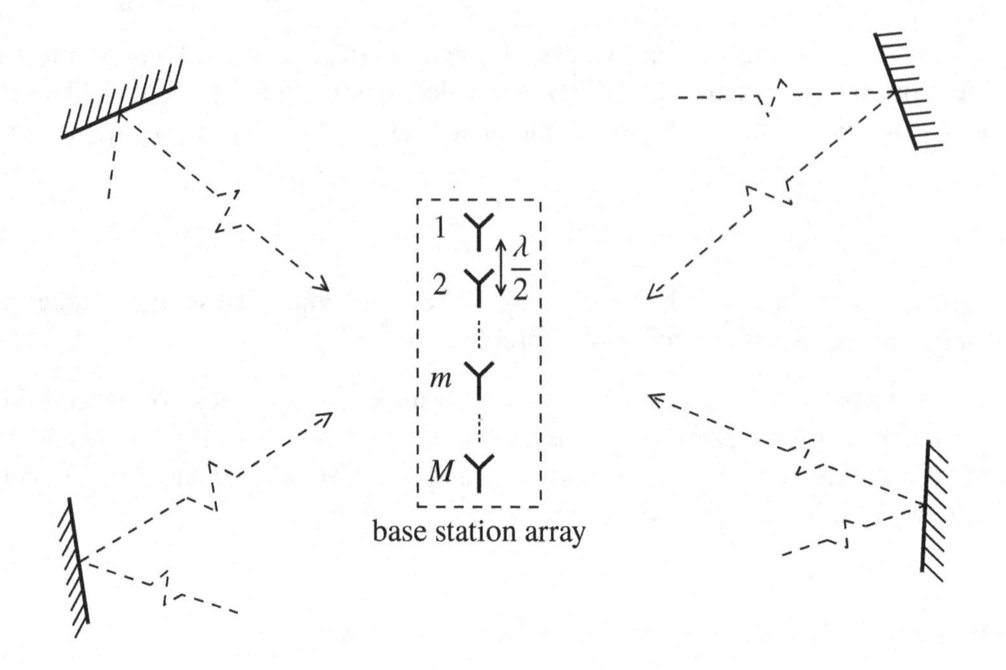

Figure 7.1. A $\lambda/2$-spaced uniform linear array located in a propagation environment with isotropic scattering, approximated by independent Rayleigh fading.

the wave equation. If one requires the random process to satisfy the wave equation, the closest model to i.i.d. has the spatial correlation function $r(d) = \text{sinc}(2d/\lambda)$, where d is the spatial separation of the two locations at which the field is sampled [37]. Since $r(d) = 0$ when d is a nonzero integer multiple of $\lambda/2$, samples taken along a straight line at a spacing of $\lambda/2$ are uncorrelated. Hence, the fading is independent between different elements in the antenna array. Furthermore, $r(d)$ is small if $d \gg \lambda$. Hence, the channel responses associated with different terminals are also mutually independent.

In independent Rayleigh fading, $\mathsf{E}\left\{|g_k^m|^2\right\} = \beta_k$ and $\mathsf{E}\left\{g_k^{m*} g_{k'}^{m'}\right\} = 0$ when $k' \neq k$ or $m' \neq m$. By the law of large numbers,

$$\frac{1}{M}\|\boldsymbol{g}_k\|^2 \to \beta_k, \qquad M \to \infty, \qquad k = 1, \ldots, K, \tag{7.14}$$

$$\frac{1}{M}\boldsymbol{g}_k^{\mathrm{H}}\boldsymbol{g}_{k'} \to 0, \qquad M \to \infty, \qquad k \neq k'. \tag{7.15}$$

Hence, in independent Rayleigh fading, we have asymptotically favorable propagation. In Section 7.2.3, we investigate to what extent we can expect the propagation to be approximately favorable for finite M.

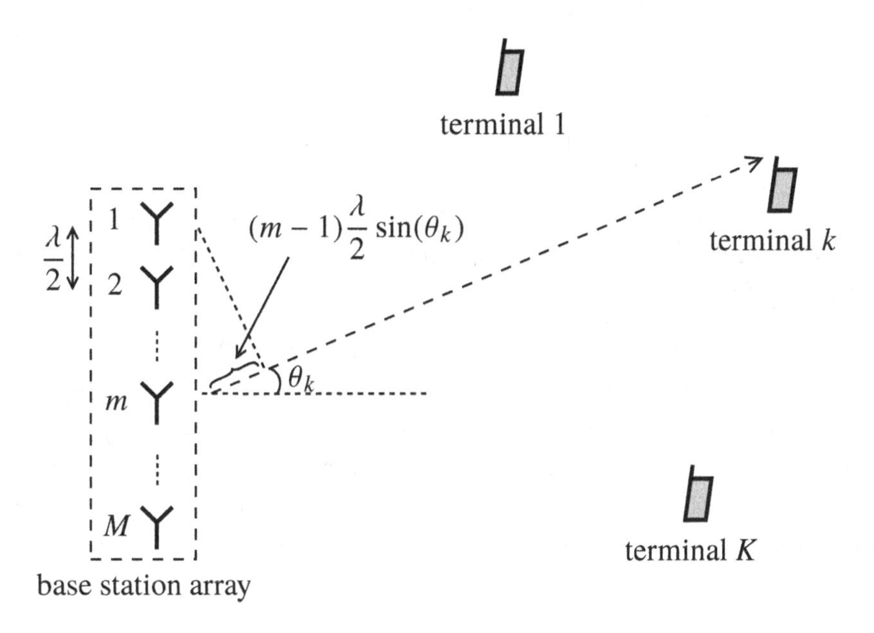

Figure 7.2. A $\lambda/2$-spaced uniform linear array located in a line-of-sight propagation environment.

7.2.2 Uniformly Random Line-of-Sight (UR-LoS)

The second scenario of interest is when there is no local scattering, and all terminals have line of sight to the base station array; see Figure 7.2. We assume that the kth terminal is located in the far-field of the array at the angle θ_k as measured relative to the array boresight. Then,

$$g_k = \sqrt{\beta_k} e^{i\phi_k} \left[1 \quad e^{-i\pi \sin(\theta_k)} \quad \ldots \quad e^{-i(M-1)\pi \sin(\theta_k)} \right]^{\mathrm{T}}, \tag{7.16}$$

where ϕ_k is a uniformly distributed random number between $-\pi$ and π that models the phase shift associated with a random range between the array and the kth terminal.

For any two terminals k and k' with angles θ_k and $\theta_{k'}$ such that $\theta_k \neq \theta_{k'}$,

$$\begin{aligned}
\frac{1}{M} g_k^{\mathrm{H}} g_{k'} &= \frac{1}{M} \sqrt{\beta_k \beta_{k'}} e^{-i(\phi_k - \phi_{k'})} \sum_{m=0}^{M-1} e^{im\pi\left(\sin(\theta_k) - \sin(\theta_{k'})\right)} \\
&= \frac{1}{M} \sqrt{\beta_k \beta_{k'}} e^{-i(\phi_k - \phi_{k'})} \frac{1 - e^{iM\pi\left(\sin(\theta_k) - \sin(\theta_{k'})\right)}}{1 - e^{i\pi\left(\sin(\theta_k) - \sin(\theta_{k'})\right)}} \\
&\to 0. \qquad M \to \infty,
\end{aligned} \tag{7.17}$$

Also,

$$\frac{1}{M}\|\boldsymbol{g}_k\|^2 = \beta_k, \qquad k = 1, \ldots, K. \tag{7.18}$$

Hence, we have asymptotically favorable propagation whenever $\{\theta_k\}$ are distinct. In Section 7.2.3, we address the question of the extent to which we can expect the propagation to be approximately favorable for finite M.

Henceforth, when we discuss line-of-sight propagation, we will assume that the terminal locations are random such that the K sin-angles $\{\sin(\theta_k)\}$ are uniformly distributed in the interval $[-1, 1]$. We refer to this assumption as *uniformly random line-of-sight* (UR-LoS). We make this assumption here for analytical convenience only, and a uniform distribution of $\{\theta_k\}$ (instead of $\{\sin(\theta_k)\}$) over $[-\pi, \pi]$ may be more realistic. However, typical antennas (such as half-wavelength patch antennas) have a directional response that discriminates against large angles of arrival, which is the regime where these two models would differ the most.

7.2.3 Independent Rayleigh Fading versus UR-LoS

For simplicity of analysis, we assume that $\beta_k = 1$ for all k. Independent Rayleigh fading can then be referred to as i.i.d. Rayleigh fading. The conclusions here can then be extrapolated, qualitatively, to general independent Rayleigh fading by re-normalizing the channel vectors $\{\boldsymbol{g}_k\}$.

In order to compare i.i.d. Rayleigh fading and UR-LoS for finite M, we first study the moments and limits of pairwise inner products $\boldsymbol{g}_k^{\mathrm{H}}\boldsymbol{g}_{k'}$ of channel vectors $\boldsymbol{g}_k$ and $\boldsymbol{g}_{k'}$. Table 7.1 summarizes the moments and limits of these inner products, and, additionally, the asymptotic ratio between the largest and smallest singular values, known from [38]. Inspection of Table 7.1 reveals that all moments and limits are the same in i.i.d. Rayleigh fading and UR-LoS, except for $\mathsf{Var}\left\{|\boldsymbol{g}_k^{\mathrm{H}}\boldsymbol{g}_{k'}|^2\right\}$, which is of the order of M times larger for UR-LoS than for i.i.d. Rayleigh fading. Hence, we expect that for finite M, i.i.d. Rayleigh fading will yield favorable channels more often than UR-LoS. Note that $\|\boldsymbol{g}_k\|^2/M \to 1$ as $M \to \infty$, in both propagation environments. This means that a coherent beamforming gain proportional to M is always achieved, and that channel hardening holds.

Recall that the notion of ergodic capacity is associated with coding over many independent realizations of all sources of randomness, and it permits us to average over $\boldsymbol{G}$ to obtain capacity expressions. This condition is typically fulfilled in the case of Rayleigh fading through the frequency-dependence of small-scale fading, and the possibility of performing coding over more than one temporal coherence interval. In UR-LoS, however, ergodic

	i.i.d. Rayleigh	UR-LoS
$\frac{1}{M}\mathsf{E}\left\{\lVert\mathbf{g}_k\rVert^2\right\}$	1	1
$\frac{1}{M}\mathsf{E}\left\{\mathbf{g}_k^{\mathrm{H}}\mathbf{g}_{k'}\right\}, \quad k \neq k'$	0	0
$\frac{1}{M}\mathsf{E}\left\{\lvert\mathbf{g}_k^{\mathrm{H}}\mathbf{g}_{k'}\rvert^2\right\}, \quad k \neq k'$	1	1
$\frac{1}{M^2}\mathsf{Var}\left\{\lvert\mathbf{g}_k^{\mathrm{H}}\mathbf{g}_{k'}\rvert^2\right\}, \quad k \neq k'$	$\frac{M+2}{M} \approx 1$	$\frac{(M-1)M(2M-1)}{3M^2} \approx \frac{2}{3}M$
$\frac{1}{M}\lVert\mathbf{g}_k\rVert^2, \quad M \to \infty$	1	1
$\frac{1}{M}\lvert\mathbf{g}_k^{\mathrm{H}}\mathbf{g}_{k'}\rvert, \quad M \to \infty, \quad k \neq k'$	0	0
$\frac{\sigma_{\max}}{\sigma_{\min}}$	$\approx \frac{1+\sqrt{\frac{K}{M}}}{1-\sqrt{\frac{K}{M}}}$	N/A

Table 7.1. Comparison of asymptotic properties and moments of pairwise channel vector inner products in i.i.d. Rayleigh and UR-LoS propagation. Here $\beta_k = 1$ for all k.

capacity is unobtainable since the random angle of arrival is independent of frequency, and remains substantially constant for long intervals of time.

Singular Value Comparisons

To provide further insight, we consider the distribution of the squared singular values $\{\sigma_k^2\}$. Recall from (7.11) and (7.12) that the sum capacity and the sum of the mean-square errors are functions of $\{\sigma_k^2\}$.

Figure 7.3 shows cumulative distribution functions of $\{\sigma_k^2\}$ in i.i.d. Rayleigh fading and UR-LoS, for two cases: $M = 100$ and $K = 10$ respectively $M = 500$ and $K = 50$. In all cases, $\beta_k = 1$. In i.i.d. Rayleigh fading, $\{\sigma_k^2\}$ are almost uniformly spread out between $\sigma_{\min}^2$ and $\sigma_{\max}^2$, and the curves have no significant tails. Since the ratio $\sigma_{\max}^2/\sigma_{\min}^2$ is small if $M \gg K$ (see Table 7.1), propagation is approximately favorable. By way of contrast, in UR-LoS a few of the singular values are very small with substantial probability while the rest are highly concentrated around their medians. This suggests that by dropping a few terminals from service in each coherence interval, UR-LoS will offer favorable propagation with very high probability. In the next section, we give a more exact argument.

Urns-and-Balls Model for UR-LoS

To quantify approximately how many terminals must be dropped from service in each coherence interval in order to have favorable propagation with high probability in the UR-LoS case, we consider the following "urns-and-balls model" [39]. A uniform linear array with M elements at inter-element spacing $\lambda/2$ can create M orthogonal beams with response vectors $\{\boldsymbol{g}_k\}$ given by (7.16) and $\{\theta_k\}$ satisfying

$$\sin(\theta_k) = -1 + \frac{2k-1}{M}, \qquad k = 1, 2, \ldots, M. \tag{7.19}$$

These vectors $\{\boldsymbol{g}_k\}$ satisfy (7.1). In terms of the angular decomposition model of MIMO channels [40] and [29, Chapter 7], the angles $\{\theta_k\}$ are *resolvable*, in the sense that sources with the angles-of-arrival $\{\theta_k\}$ have distinct spatial signatures.

Suppose that each one of the K terminals is associated with one of the possible angles $\{\theta_k\}$ defined implicitly by (7.19). This means that each terminal is randomly and independently assigned to one of M orthogonal beams; see Figure 7.4. For the channel to offer approximately favorable propagation, each of the M beams must contain at most one terminal. If there are two or more terminals in a given beam, all but one of those terminals have to be dropped from service in a given coherence interval in order for the propagation to be favorable. In case some beams contain two or more terminals and none of them is dropped, the remaining terminals in all other beams – that are occupied by at most one terminal – would still experience favorable propagation.

Under the assumptions that underlie the UR-LoS model, each of the K terminals is equally likely to fall into any of the M beams. Let N_0 be the number of beams that have no terminal. Then, $M - K \leq N_0 < M$, and the number of terminals that must be dropped from service is $N_{\text{drop}} = K - (M - N_0)$. The probability that n terminals, $0 \leq n < K$, are dropped is given by

$$\begin{aligned} \mathrm{P}\left(N_{\text{drop}} = n\right) &= \mathrm{P}(K - (M - N_0) = n) \\ &= \mathrm{P}(N_0 = n + M - K) \\ &\overset{(a)}{=} \binom{M}{n+M-K} \sum_{k=1}^{K-n} (-1)^k \binom{K-n}{k} \left(1 - \frac{n+M-K+k}{M}\right)^K, \end{aligned} \tag{7.20}$$

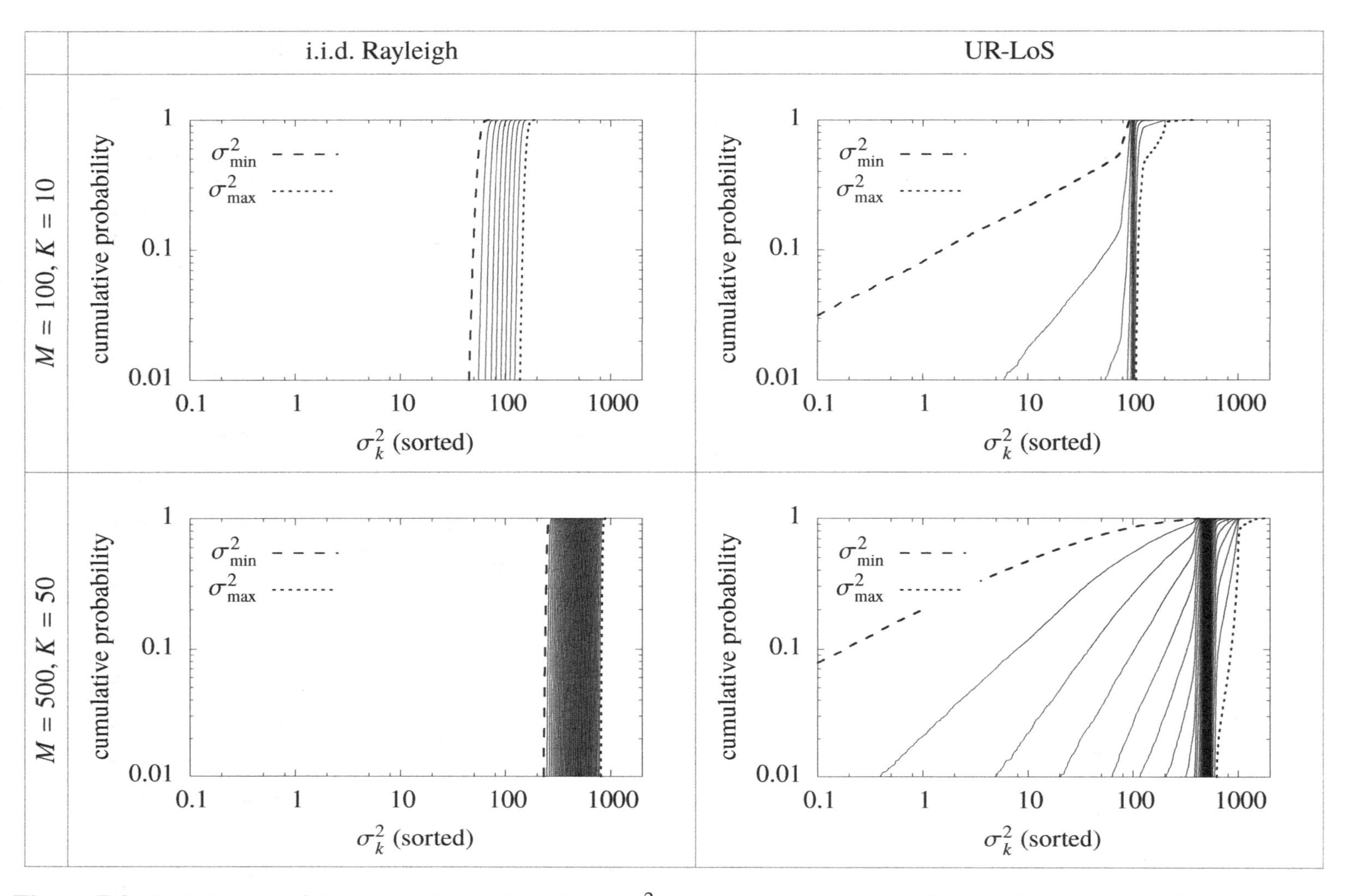

Figure 7.3. Distribution of the squared singular values $\{\sigma_k^2\}$ of $\boldsymbol{G}$ for i.i.d. Rayleigh fading ($\beta_k = 1$ for all k) and UR-LoS propagation.

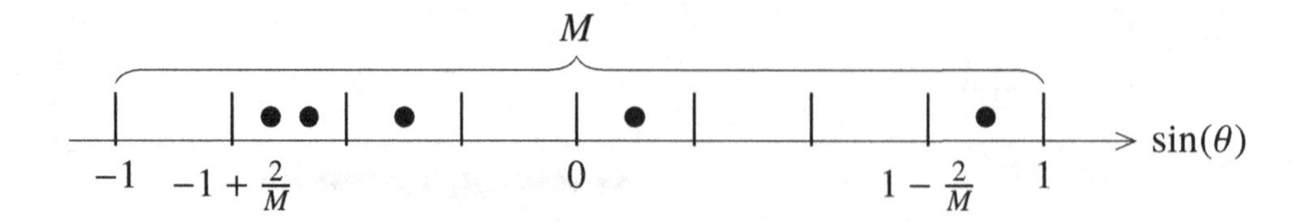

Figure 7.4. Urns-and-balls model for UR-LoS propagation.

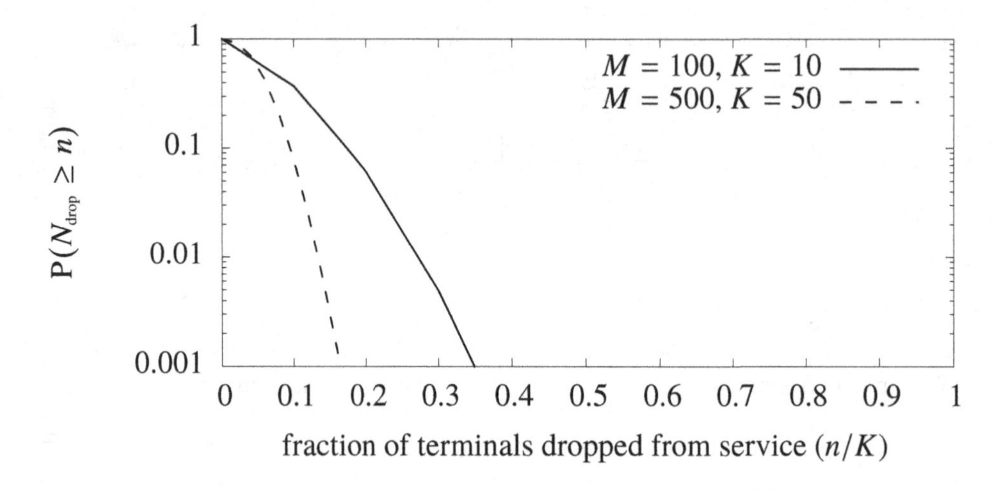

Figure 7.5. The probability that n or more terminals must be dropped from service in each coherence interval, using the urns-and-balls model for UR-LoS propagation.

where (a) follows from a standard combinatorial argument [41, Eq. (2.4)].

Figure 7.5 shows the probability that n or more terminals must be dropped from service,

$$\mathrm{P}\left(N_{\text{drop}} \geq n\right) = \sum_{n'=n}^{K-1} \mathrm{P}\left(N_{\text{drop}} = n'\right), \tag{7.21}$$

for two cases: $M = 100, K = 10$ and $M = 500, K = 50$. In the case of $M = 100$ and $K = 10$, the probability that three or more terminals have to be dropped in each coherence interval is about 1%. Similarly, in the case of $M = 500$ and $K = 50$, the probability that eight or more terminals have to be dropped is about 1%. This is consistent with the intuition developed from Figure 7.3: the presence of a few very small singular values with appreciable probability suggests that a few terminals must be dropped from service in every coherence interval.

Capacity Comparisons

As a further quantitative example, Figure 7.6 compares the cumulative distribution function of the uplink capacity per terminal – that is, the uplink sum capacity C_{sum} in (7.3) divided by K, and the corresponding favorable propagation upper bound (7.4) divided by K. The ratio between these two quantities is Δ_C in (7.5). The figures display capacities for $M = 100$ and $K = 10$ under both i.i.d. Rayleigh fading and UR-LOS conditions when all ten terminals are served and when eight selected terminals are served.

When all ten terminals are considered, the actual capacity is very close to its upper bound with high probability in i.i.d. Rayleigh fading, but not in UR-LoS. The urns-and-balls model analysis suggests that by dropping two terminals from service, the remaining terminals will experience favorable propagation. As we see in Figure 7.6, this is indeed the case: by choosing 8 out of 10 terminals that give the highest sum capacity, the sum capacity is very close to the upper bound also in the UR-LoS case.

The results of Figure 7.6 are concerned with the arithmetic mean of the per-terminal spectral efficiency. Substantially the same conclusions are observed, for each realization of the small-scale fading, if linear processing and max-min power control are performed.

Discussion

Both independent Rayleigh fading and UR-LoS environments offer approximately favorable propagation, provided that a few terminals can be dropped from service in every coherence interval in the UR-LoS case. These two scenarios represent rather extreme cases, and in reality we are more likely to have a situation in between these two cases. It is then reasonable to expect that in most practical cases we have favorable propagation to a large extent. This conclusion has been confirmed experimentally by several independent measurement campaigns [42–46], even though the particular topology of the array can play a significant role [47].

Throughout the analysis, we have assumed that the base station array is uniform and linear with an inter-element spacing of $\lambda/2$. With a different array, the above conclusions may or may not hold precisely as stated. Consider, as a first example, a uniform linear array with twice the inter-element spacing, λ. In isotropic scattering, the correlation function $r(d) = \text{sinc}(2d/\lambda)$ equals zero when d is an integer multiple of $\lambda/2$; hence the fading is uncorrelated between the antenna elements and we have independent Rayleigh fading, as in the case with inter-element spacing $\lambda/2$. In UR-LoS, the array still offers M "beams," although each beam now is split into two sub-beams with half the angular width [40]. The urns-and-balls model can still be used to analyze this situation, and the result is substantially

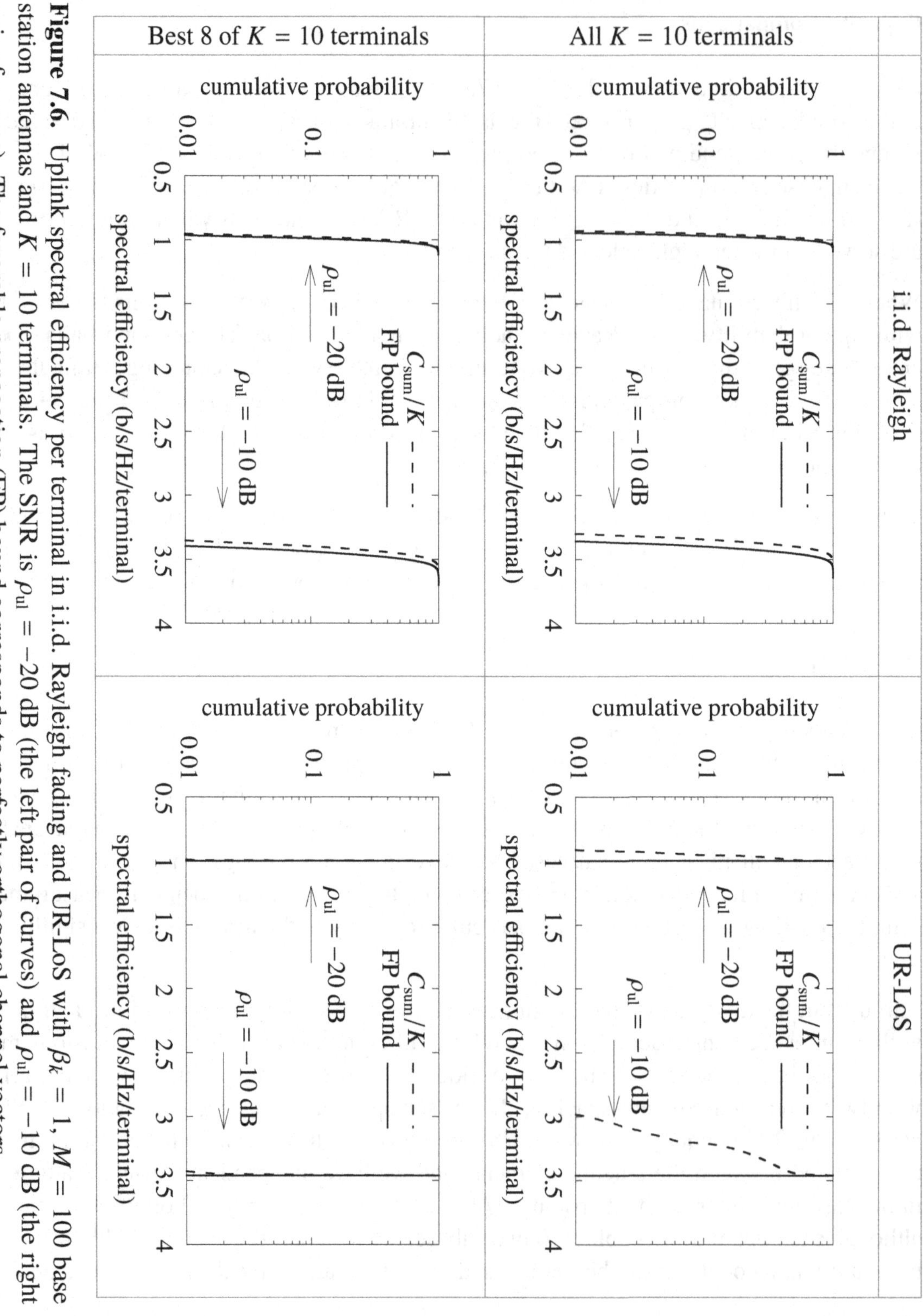

Figure 7.6. Uplink spectral efficiency per terminal in i.i.d. Rayleigh fading and UR-LoS with $\beta_k = 1$, $M = 100$ base station antennas and $K = 10$ terminals. The SNR is $\rho_{ul} = -20$ dB (the left pair of curves) and $\rho_{ul} = -10$ dB (the right pair of curves). The favorable propagation (FP) bound corresponds to perfectly orthogonal channel vectors.

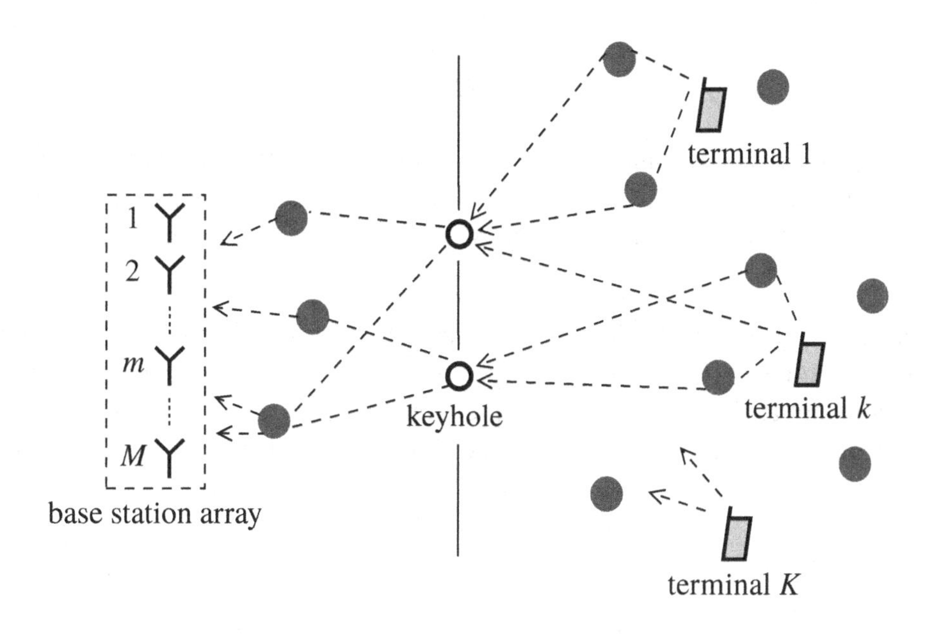

Figure 7.7. Keyhole channel with $N = 2$ keyholes.

the same as in the case of an antenna spacing of $\lambda/2$.

As a second example, consider a uniform *rectangular* array for which each element has its closest neighbor at a distance of $\lambda/2$. In this case, the distance between diagonally neighboring elements is $\lambda/\sqrt{2}$, and the fading is now correlated between two such elements. Despite this mild correlation, numerical experiments disclose that capacity is not significantly reduced from the i.i.d. case.

7.3 Finite-Dimensional Channels

Independent Rayleigh fading and UR-LoS are two examples of practical situations where we expect propagation to be favorable. In contrast, Figure 7.7 shows a situation that may arise in practice and where the propagation is *not* likely to be favorable. Here, all paths from the base station array to the terminals pass through N *keyholes* – depicted as slots through a wall in the figure, which is small compared to K; $N < K$.

Here, the $M \times K$ channel matrix $\boldsymbol{G}$ is rank deficient since it can be written as the product of an $M \times N$ matrix $\boldsymbol{G}_a$ that models the channel between the keyholes and the base station array, and an $N \times K$ matrix $\boldsymbol{G}_b$ that describes the channel between the terminals and the keyholes,

$$G = G_a G_b. \tag{7.22}$$

The rank of the product $G_a G_b$ cannot exceed the rank of G_a, which in turn cannot exceed the number of keyholes, N. Hence the rank of G cannot be larger than N. Therefore, (7.1) cannot be satisfied, irrespective of how large M is. Propagation can only be favorable if $N \geq K$.

Another effect that will be visible in the environment illustrated in Figure 7.7 is that the channel may not harden when $M \to \infty$. Specifically, $\|g_k\|^2/M$ may not converge to a deterministic constant as $M \to \infty$. For example, consider a channel with a single keyhole and a single terminal; $N = 1$ and $K = 1$. Then, dropping the index $k = 1$ for simplicity, we have

$$g = \sqrt{\beta} g_a g_b, \tag{7.23}$$

where g_a is an $M \times 1$ random vector and g_b is a random scalar. The elements of g_a and the scalar g_b are well modeled as i.i.d. $CN(0, 1)$ random variables. As $M \to \infty$, $\|g_a\|^2/M \to 1$, but $\|g\|^2/M$ does not converge to a deterministic constant. Hence, the channel does not harden and the properties of g are very different from those in independent Rayleigh fading.

7.4 Summary of Key Points

- Propagation is said to be (approximately) *favorable* if the channel responses $\{g_k\}$ in a given cell are pairwisely (nearly) orthogonal.

  Favorable propagation brings several benefits. For example, on the uplink, the sum capacity can be upper-bounded by a value that depends only on the channel norms $\{\|g_k\|^2\}$, and a channel that offers favorable propagation achieves this upper bound. This means that there are no other channels $\{g_k\}$ with the same norms $\{\|g_k\|^2\}$ that offer a higher sum capacity. Also, the error variance with linear detection is minimized if propagation is favorable.

- Two environments that offer approximately favorable propagation are isotropic scattering resulting in independent Rayleigh fading (see Section 7.2.1), and Uniformly Random Line-of-Sight (see Section 7.2.2). In UR-LoS, a few terminals have to be dropped from service in every coherence interval in order to guarantee favorable propagation. The urns-and-balls model of Section 7.2.3 represents a useful way of thinking about the need to drop terminals from service in UR-LoS propagation.

- The switch from isotropic scattering to UR-LoS does not entail significant reduction in performance, a fact that should lend considerable confidence to performance analyses based on i.i.d. Rayleigh fading.

Chapter 8

FINAL NOTES AND FUTURE DIRECTIONS

This chapter summarizes some open questions that demand further research, and briefly discusses selected topics that we have been unable to cover in the book but on which substantial investigations have taken place, or are taking place at the time of this writing.

8.1 Alternative Approaches to Performance Analysis

Our capacity analyses use techniques first developed in [48] for the uplink, and [49] for the downlink. This approach has several advantages: the resulting capacity bounds are rigorous and do not involve approximations, and the resulting expressions are in closed form and easy to interpret intuitively.

Alternative approaches to the capacity analysis of Massive MIMO exist. Most notably, the sequence of papers [50–52] used asymptotic random-matrix theoretic results in order to obtain "deterministic equivalent" capacity expressions. An advantage of that approach is that MMSE and regularized zero-forcing processing can be analyzed, whereas the theory developed in this book is only applicable to zero-forcing and maximum-ratio processing. While our capacity analyses assume independent small-scale fading, [50–52] accounted for spatial correlation; however, they were unable to obtain exact closed-form performance expressions.

8.2 Multiple-Antenna Terminals

Massive MIMO, in contrast to Point-to-Point MIMO, works extremely well with only single-antenna terminals. Nothing in our discussion, however, precludes the use of multiple-antenna terminals. A multiple-antenna terminal, for example, could enjoy throughput in proportion to the number of antennas that it possesses without requiring exponentially growing SINRs. The simplest mode of operation would treat the multiple-antenna terminal as a multiplicity of single-antenna terminals. Alternatively, a terminal could use multiple antennas for out-of-system interference suppression, facilitated by adding a "quiet interval" in each coherence interval, during which none of the base stations or terminals transmits. The terminals (and incidentally the base station) could each identify the subspace in which the interference is contained, and then operate in the orthogonal interference-free subspace.

8.3 Pilot Contamination and Pilot Assignment

We saw in Chapter 6 that pilot contamination can be a significant impairment in a multi-cell system and that it may be mitigated by adopting a pilot reuse factor greater than one. However, the greater overhead of high reuse factors limits the number of mobile terminals that can simultaneously be served.

Is pilot contamination an unavoidable fundamental limitation in a multi-cell system, or is it an artifact of suboptimal signal processing, in particular MMSE channel estimation and linear processing? With MMSE channel estimation and linear processing, the consequence of pilot contamination is that all known capacity lower bounds approach a finite limit when $M \to \infty$, as shown in Section 4.4.1 and in several papers [48, 53]. This is true even under more specific model assumptions – for example, if the channel is finite-dimensional [54]. It is also known that contamination by payload data has similar effects as contamination by pilots in other cells; see Section 4.4.3 and [55].

In contrast, we do not know of any relevant, non-trivial *upper* bounds on capacity that account for the effects of imperfect channel knowledge. It is known, however, for the related cases of Point-to-Point MIMO [11] and cellular systems with base station cooperation [56] that capacity is ultimately limited by the availability of sufficiently accurate CSI, which in turn is limited by the length of the channel coherence interval. Hence, we conjecture that pilot contamination in Massive MIMO is an inevitable limitation that cannot be entirely overcome. Nevertheless, sophisticated pilot assignment and signal processing may significantly reduce the problem.

Throughout the book, we considered universal reuse of resources for the payload data, but potentially sparser reuse of the pilots. Adaptation and optimization of the pilot allocation between cells can further reduce the effects of pilot contamination [57–64]. Difference in spatial correlation may be exploited to reuse pilots without incurring appreciable contamination effects [65–67].

Several papers [68–72] proposed non-linear channel estimation algorithms that exploit both received pilots and uplink payload data, and which can mitigate the pilot contamination problem to some extent. References [73–76] developed techniques to eliminate pilot contamination by joint, coherent processing among base stations. These methods require payload data and large-scale fading coefficients to be shared among the base stations, but no small-scale CSI is shared.

8.4 Massive MIMO with FDD Operation

We have consistently assumed that Massive MIMO operates in TDD mode rather than in FDD mode. As explained in Section 1.4, TDD is fundamentally superior to FDD. However, for various commercial reasons there remains substantial interest in finding ways of making Massive MIMO work in FDD. Subject to the assumption of enough special structure in the propagation, the performance of FDD operation may be comparable to that of TDD operation [77].

FDD-based Massive MIMO requires feedback of channel estimates from the terminals to the base station. The references [78–80] proposed efficient non-coherent trellis quantization and vector quantization methods for the feedback, and associated optimized downlink training schemes. Other papers exploited assumed sparsity of the channel response in a particular domain in order to reduce the quantity of downlink pilots and uplink feedback [81, 82].

The papers [83,84] proposed a type of two-stage precoding: an outer precoding for inter-cell and inter-cluster interference cancellation, and an inner precoding for intra-cell spatial multiplexing. Reference [85] proposed a similar dual-structured precoding for Massive MIMO systems with multi-polarized antenna elements. In [86], a transmission scheme for Massive MIMO that requires only statistical CSI at the base station was proposed.

8.5 Cell-Free Massive MIMO

Massive MIMO systems can be envisioned where the base station antennas are distributed over a large geographical area, rather than co-located in a compact array. The following discussion refers to such configurations as *cell-free*. In a cell-free deployment, the large-scale fading coefficients depend on both the antenna index and the terminal index.

A number of research papers studied the performance of cell-free Massive MIMO systems under the assumption that perfect CSI is available everywhere [87–91]. The paper [92] compared the downlink performance of cell-free Massive MIMO with that of Massive MIMO using co-located arrays and zero-forcing processing, also under the assumption of perfect CSI.

A few papers have studied the performance of cell-free Massive MIMO with imperfect CSI. Reference [93] derived approximate expressions for the uplink performance with maximum-ratio and MMSE processing, taking into account the effects of channel estimation but assuming that all pilot sequences are orthogonal. In [94], rigorous capacity lower bounds for maximum-ratio processing and algorithms for pilot allocation and max-min SINR power control were obtained, accounting for channel estimation errors and assuming pilot reuse – and therefore pilot contamination – within the network. Hybrid architectures are also possible where base stations use large antenna arrays which operate cooperatively [66, 95].

8.6 Signal Processing Algorithms

Linear signal processing is sufficient to reach the capacity lower bounds in Chapters 3 and 4. Under certain conditions, more refined algorithms may yield improvements.

In the most elementary form of Massive MIMO, as presented in this book, the terminals do not attempt to acquire CSI. In some cases, the downlink performance can be increased by having every terminal estimate its effective (scalar) channel gain, either through downlink pilots transmitted through the beams [96,97], or by using to blind estimation techniques [98].

Implementation of zero-forcing requires the computation of matrix inversions or solutions of linear systems of equations. To that end, computationally efficient approximations based on truncated polynomial expansions [99–102], or conjugate-gradient techniques [103], are available.

More advanced precoders can be advantageous. References [104–106] developed precoding for hybrid transceiver architectures that dispenses with the RF chain of each antenna, and that instead uses a steerable phase-shifting network inserted between the RF frontend and the antennas. For distributed Massive MIMO systems, [107] proposed a precoding technique with individual CSI.

In the simpler forms of Massive MIMO, only K out of M spatial degrees of freedom are employed. The $M - K$ unused degrees of freedom can be exploited in several ways. For example, when the pilot reuse factor is greater than one, $n_{\text{reuse}} > 1$, each base station can learn the channels both to the terminals in its own cell and to the terminals in neighboring cells that use distinct pilots. The obtained channel estimates can then be used to suppress inter-cell interference both in the uplink and the downlink [61, 108, 109]. Another possible use of the spare degrees of freedom is on the downlink to place signal components in the nullspace of the channel matrix. Within the limitations imposed by channel estimation error, these components are invisible to the terminals receiving downlink data. Several papers [110–113] have proposed to add invisible signal components in such a way that the waveforms emitted by each antenna have low peak-to-average-ratios. Other possible uses of the spare degrees of freedom include broadcasting of information to terminals that are not targeted by the beamforming [114], and transmission of artificial noise to improve secrecy rates [115].

8.7 Effects of Non-Ideal Hardware

Massive MIMO exploits the law of large numbers when $M \gg 1$ to average out the effects of small-scale fading, receiver noise, and non-coherent interference. There is some evidence that distortion arising from hardware impairments, including non-linearities, I/Q-imbalance, and phase noise, also averages out to a substantial extent.

The papers [116–118] analyzed the impact of in-band distortion from non-ideal transceiver hardware, and showed that when M is large, the hardware quality of the terminals is typically the limiting factor. There is a body of work that deals with specific types of hardware impairments, for example I/Q imbalance [119, 120], quantization noise from low-resolution analog-to-digital converters [121], power amplifier non-linearities [122], and phase noise [117, 123–125].

Massive MIMO with TDD operation requires that the uplink and downlink channels be reciprocal, as discussed in Chapter 1. In practice, reciprocity calibration of the hardware chains is necessary. Algorithms that accomplish such calibration can be found in [126, 127] and in [128–132] for Massive MIMO specifically. These solutions either rely on extra antennas dedicated to the calibration, or two-way channel measurements between base stations and terminals, or on mutual coupling within the base station antenna array. Analyses of the impact of reciprocity mismatch are contained in [133, 134].

8.8 Random Access and Resource Allocation

Massive MIMO creates virtual circuits between the serving base station and K terminals in the cell, characterized by the functional block diagrams in Figures 3.1–3.4 and 4.1–4.4. To achieve a high net sum throughput in a cell, a substantial number of terminals need to be multiplexed simultaneously. This in turn requires appropriate algorithms for scheduling, random access, and dynamic pilot assignment. These problems occur in conventional Multiuser MIMO, but are magnified by the scale of Massive MIMO. Some initial approaches were reported in [135, 136]. Internet-of-Things, machine-to-machine communications, and sensor networks represent possible applications of Massive MIMO for which scheduling would be complicated by sporadic service requirements. In [137], such scenarios were considered, taking into account intermittent terminal activity and possible pilot collisions within the cell.

Practical Massive MIMO systems may require the base station to transmit information to terminals for which it has not yet acquired CSI. This information may include control signaling or other public information. In the absence of CSI, beamforming is impossible and the only way of benefitting from multiple antennas is to use space-time coding, which does not offer multiplexing or array gains. Space-time coding in turn requires either the transmission of downlink pilots or non-coherent communication. Some results on the tradeoffs involved in the selection of appropriate codes here can be found in [138, 139].

Throughout this book, we have assumed that pilots are always transmitted with full power, whereas power control is applied to the payload data. In some cases, adjusting the power of the pilots can improve performance [140].

8.9 Total Energy Efficiency

The large array gain offered by Massive MIMO translates into substantial savings in radiated power. In some deployment scenarios, the energy efficiency of wireless networks is a concern. The total energy consumption, which includes both radiated power and power dissipated in circuits and spent on signal processing, is then a relevant measure. A body of literature exists that gives insights into the total energy-efficiency aspects of Massive MIMO [141–145].

Appendix A

CIRCULARLY SYMMETRIC COMPLEX GAUSSIAN RANDOM VECTORS

This appendix reviews some elementary properties of circularly symmetric complex Gaussian random vectors, which are used throughout the book.

A.1 Complex Gaussian Random Vectors

Baseband quantities are naturally complex-valued. In general, a complete characterization of a complex random vector entails a specification of the joint distribution of its real and imaginary parts. If z' and z'' are $M \times 1$ real-valued, jointly Gaussian random vectors, then

$$z = z' + iz'' \tag{A.1}$$

is a complex Gaussian vector. The probability density function of z is the joint probability density function of its real and imaginary parts, that is, of the $2M$ real-valued Gaussian random variables that constitute z.

In general, the fact that z is complex-valued does not lend any simplification to the specification of its probability density. However the special property of circular symmetry does permit considerable simplifications.

A.2 Circularly Symmetric Complex Gaussian Random Vectors

Let z be an $M \times 1$ complex-valued Gaussian vector. Then z is said to be *circularly symmetric* if $e^{i\phi}z$ has the same probability distribution as z for all real-valued ϕ. Note that if z is circularly symmetric, then for any ϕ we have that

$$\mathsf{E}\left\{e^{i\phi}z\right\} = \mathsf{E}\{z\}, \tag{A.2}$$

so z must have zero mean:

$$\mathsf{E}\{z\} = \mathbf{0}. \tag{A.3}$$

If z is circularly symmetric Gaussian (with zero mean), then we write for short,

$$z \sim \mathrm{CN}(\mathbf{0}, \boldsymbol{\Lambda}), \tag{A.4}$$

where $\boldsymbol{\Lambda}$ is the covariance:

$$\begin{aligned} \boldsymbol{\Lambda} &= \mathsf{Cov}\{z\} \\ &= \mathsf{E}\left\{zz^{\mathrm{H}}\right\}. \end{aligned} \tag{A.5}$$

Circularly symmetric random vectors arise naturally in the modeling of noise, where the absolute phase is random. They also naturally arise in models of wireless communications links, where transmitters and receivers are separated by hundreds of wavelengths or more, and it is perfectly accurate to treat the overall absolute phase as unknown and uniformly distributed.

Clearly, if $\boldsymbol{x}$ is an arbitrary (not necessarily circularly symmetric) zero-mean, complex Gaussian vector, and we construct a vector

$$z = \boldsymbol{x}e^{i\theta}, \tag{A.6}$$

where θ is a uniformly distributed phase, then $e^{i\phi}z$ has the same distribution as z for all real-valued ϕ, so z is circularly symmetric.

A.2.1 Pseudo-Covariance Matrix

If z is circularly symmetric, then for any real-valued ϕ,

$$\begin{aligned} \mathsf{E}\left\{zz^{\mathrm{T}}\right\} &= \mathsf{E}\left\{e^{i\phi}z\left(e^{i\phi}z\right)^{\mathrm{T}}\right\} \\ &= e^{i2\phi}\mathsf{E}\left\{zz^{\mathrm{T}}\right\}. \end{aligned} \tag{A.7}$$

Hence,

$$\mathsf{E}\left\{zz^{\mathrm{T}}\right\} = \mathbf{0}. \tag{A.8}$$

The matrix $\mathsf{E}\left\{zz^{\mathrm{T}}\right\}$ is called the *pseudo-covariance matrix* of z.

Conversely, if z is a complex Gaussian random vector and (A.8) holds, then we show in the next section that z must be circularly symmetric.

A.2.2 Probability Density Function

Let z be an $M \times 1$ circularly symmetric complex Gaussian vector, with its real and imaginary parts defined as in (A.1). The covariance of z can be decomposed as follows:

$$\boldsymbol{\Lambda} = \boldsymbol{\Lambda}' + i\boldsymbol{\Lambda}'', \tag{A.9}$$

where

$$\boldsymbol{\Lambda}' = \mathsf{E}\left\{z'z'^{\mathrm{T}}\right\} + \mathsf{E}\left\{z''z''^{\mathrm{T}}\right\}, \tag{A.10}$$

$$\boldsymbol{\Lambda}'' = -\mathsf{E}\left\{z'z''^{\mathrm{T}}\right\} + \mathsf{E}\left\{z''z'^{\mathrm{T}}\right\}. \tag{A.11}$$

Owing to the circular symmetry,

$$\begin{aligned}\mathsf{E}\left\{zz^{\mathrm{T}}\right\} &= \left(\mathsf{E}\left\{z'z'^{\mathrm{T}}\right\} - \mathsf{E}\left\{z''z''^{\mathrm{T}}\right\}\right) + i\left(\mathsf{E}\left\{z'z''^{\mathrm{T}}\right\} + \mathsf{E}\left\{z''z'^{\mathrm{T}}\right\}\right)\\ &= \mathbf{0}.\end{aligned} \tag{A.12}$$

Therefore,

$$\begin{aligned}\mathsf{E}\left\{z'z'^{\mathrm{T}}\right\} &= \mathsf{E}\left\{z''z''^{\mathrm{T}}\right\}, \quad \text{and}\\ \mathsf{E}\left\{z'z''^{\mathrm{T}}\right\} &= -\mathsf{E}\left\{z''z'^{\mathrm{T}}\right\}.\end{aligned} \tag{A.13}$$

Equations (A.10), (A.11), and (A.13) yield

$$\mathsf{E}\left\{\begin{bmatrix} z' \\ z'' \end{bmatrix}\begin{bmatrix} z'^{\mathrm{T}} & z''^{\mathrm{T}} \end{bmatrix}\right\} = \frac{1}{2}\begin{bmatrix} \boldsymbol{\Lambda}' & -\boldsymbol{\Lambda}'' \\ -\boldsymbol{\Lambda}''^{\mathrm{T}} & \boldsymbol{\Lambda}' \end{bmatrix}. \tag{A.14}$$

Using (A.14), the formula for the probability density of the $2M$ real-valued components of z is

$$p_z(\zeta) = \frac{\exp\left(-\frac{1}{2}\begin{bmatrix} \zeta'^{\mathrm{T}} & \zeta''^{\mathrm{T}} \end{bmatrix}\left(\frac{1}{2}\begin{bmatrix} \boldsymbol{\Lambda}' & -\boldsymbol{\Lambda}'' \\ -\boldsymbol{\Lambda}''^{\mathrm{T}} & \boldsymbol{\Lambda}' \end{bmatrix}\right)^{-1}\begin{bmatrix} \zeta' \\ \zeta'' \end{bmatrix}\right)}{(2\pi)^{2M/2}\left|\frac{1}{2}\begin{bmatrix} \boldsymbol{\Lambda}' & -\boldsymbol{\Lambda}'' \\ -\boldsymbol{\Lambda}''^{\mathrm{T}} & \boldsymbol{\Lambda}' \end{bmatrix}\right|^{1/2}}, \tag{A.15}$$

where ζ' and ζ'' are real-valued $M \times 1$-vectors and $\zeta = \zeta' + i\zeta''$.

To evaluate the determinant and the inverse of the covariance matrix in (A.15), we perform a block-diagonalization as follows:

$$\begin{aligned}&\left(\frac{1}{\sqrt{2}}\begin{bmatrix} \boldsymbol{I}_M & i\boldsymbol{I}_M \\ i\boldsymbol{I}_M & \boldsymbol{I}_M \end{bmatrix}\right)\left(\frac{1}{2}\begin{bmatrix} \boldsymbol{\Lambda}' & -\boldsymbol{\Lambda}'' \\ -\boldsymbol{\Lambda}''^{\mathrm{T}} & \boldsymbol{\Lambda}' \end{bmatrix}\right)\left(\frac{1}{\sqrt{2}}\begin{bmatrix} \boldsymbol{I}_M & -i\boldsymbol{I}_M \\ -i\boldsymbol{I}_M & \boldsymbol{I}_M \end{bmatrix}\right)\\ &\qquad = \left(\frac{1}{\sqrt{2}}\begin{bmatrix} \boldsymbol{I}_M & i\boldsymbol{I}_M \\ i\boldsymbol{I}_M & \boldsymbol{I}_M \end{bmatrix}\right)\left(\frac{1}{2\sqrt{2}}\begin{bmatrix} \boldsymbol{\Lambda} & -i\boldsymbol{\Lambda}^* \\ -i\boldsymbol{\Lambda} & \boldsymbol{\Lambda}^* \end{bmatrix}\right)\\ &\qquad = \frac{1}{2}\begin{bmatrix} \boldsymbol{\Lambda} & \mathbf{0} \\ \mathbf{0} & \boldsymbol{\Lambda}^* \end{bmatrix},\end{aligned} \tag{A.16}$$

where we have used the skew-symmetry of the imaginary part of the covariance, $\mathbf{\Lambda}''^{\mathrm{T}} = -\mathbf{\Lambda}''$. The left- and right-hand factors in (A.16) are unitary, so the determinant in (A.15) is

$$\begin{aligned} \left| \frac{1}{2} \begin{bmatrix} \mathbf{\Lambda}' & -\mathbf{\Lambda}'' \\ -\mathbf{\Lambda}''^{\mathrm{T}} & \mathbf{\Lambda}' \end{bmatrix} \right| &= \frac{1}{2^{2M}} |\mathbf{\Lambda}| \, |\mathbf{\Lambda}^*| \\ &= \frac{1}{2^{2M}} |\mathbf{\Lambda}|^2 . \end{aligned} \tag{A.17}$$

The inverse of the covariance follows directly from (A.16):

$$\begin{aligned} &\left(\frac{1}{2} \begin{bmatrix} \mathbf{\Lambda}' & -\mathbf{\Lambda}'' \\ -\mathbf{\Lambda}''^{\mathrm{T}} & \mathbf{\Lambda}' \end{bmatrix} \right)^{-1} \\ &= \left(\frac{1}{\sqrt{2}} \begin{bmatrix} \boldsymbol{I}_M & -i\boldsymbol{I}_M \\ -i\boldsymbol{I}_M & \boldsymbol{I}_M \end{bmatrix} \right) \left(2 \begin{bmatrix} \mathbf{\Lambda}^{-1} & \mathbf{0} \\ \mathbf{0} & (\mathbf{\Lambda}^*)^{-1} \end{bmatrix} \right) \left(\frac{1}{\sqrt{2}} \begin{bmatrix} \boldsymbol{I}_M & i\boldsymbol{I}_M \\ i\boldsymbol{I}_M & \boldsymbol{I}_M \end{bmatrix} \right). \end{aligned} \tag{A.18}$$

Using (A.18), the quadratic quantity in (A.15) is evaluated as follows:

$$\begin{aligned} &\frac{1}{2} \begin{bmatrix} \boldsymbol{\zeta}'^{\mathrm{T}} & \boldsymbol{\zeta}''^{\mathrm{T}} \end{bmatrix} \left(\frac{1}{2} \begin{bmatrix} \mathbf{\Lambda}' & -\mathbf{\Lambda}'' \\ -\mathbf{\Lambda}''^{\mathrm{T}} & \mathbf{\Lambda}' \end{bmatrix} \right)^{-1} \begin{bmatrix} \boldsymbol{\zeta}' \\ \boldsymbol{\zeta}'' \end{bmatrix} \\ &\qquad = \frac{1}{2} \begin{bmatrix} \boldsymbol{\zeta}^{\mathrm{H}} & -i\boldsymbol{\zeta}^{\mathrm{T}} \end{bmatrix} \begin{bmatrix} \mathbf{\Lambda}^{-1} & \mathbf{0} \\ \mathbf{0} & (\mathbf{\Lambda}^*)^{-1} \end{bmatrix} \begin{bmatrix} \boldsymbol{\zeta} \\ i\boldsymbol{\zeta}^* \end{bmatrix} \\ &\qquad = \boldsymbol{\zeta}^{\mathrm{H}} \mathbf{\Lambda}^{-1} \boldsymbol{\zeta}. \end{aligned} \tag{A.19}$$

Substitution of (A.17) and (A.19) into (A.15) yields the equivalent probability density,

$$p_{\boldsymbol{z}}(\boldsymbol{\zeta}) = \frac{1}{\pi^M |\mathbf{\Lambda}|} \exp\left(-\boldsymbol{\zeta}^{\mathrm{H}} \mathbf{\Lambda}^{-1} \boldsymbol{\zeta} \right). \tag{A.20}$$

Note from (A.20) that $\boldsymbol{z}$ and $e^{i\phi}\boldsymbol{z}$ have the same distribution for any ϕ. Hence, any random vector with the probability density in (A.20) is circularly symmetric Gaussian. It also follows that if a complex Gaussian random vector satisfies (A.8), then it must be circularly symmetric.

A.2.3 Linear Transformations

Suppose $\boldsymbol{z}$ is circularly symmetric Gaussian with covariance $\mathbf{\Lambda}$, let $\boldsymbol{A}$ be an arbitrary deterministic, complex-valued matrix, and let $\boldsymbol{y} = \boldsymbol{A}\boldsymbol{z}$. Since the real and imaginary parts of $\boldsymbol{y}$ are linear functions of the real and imaginary parts of $\boldsymbol{z}$, $\boldsymbol{y}$ is a complex Gaussian vector. Furthermore, since $\boldsymbol{z}$ is circularly symmetric, $e^{i\phi}\boldsymbol{z}$ has the same distribution as $\boldsymbol{z}$ for any ϕ. Hence, $e^{i\phi}\boldsymbol{y} = e^{i\phi}\boldsymbol{A}\boldsymbol{z}$ has the same distribution as $\boldsymbol{y}$ for any ϕ. It follows that $\boldsymbol{y}$ is circularly symmetric Gaussian. Its covariance matrix is

$$\begin{aligned}\mathsf{Cov}\{\boldsymbol{y}\} &= \mathsf{E}\left\{\boldsymbol{y}\boldsymbol{y}^{\mathrm{H}}\right\}\\ &= \mathsf{E}\left\{\boldsymbol{A}\boldsymbol{z}\boldsymbol{z}^{\mathrm{H}}\boldsymbol{A}^{\mathrm{H}}\right\}\\ &= \boldsymbol{A}\boldsymbol{\Lambda}\boldsymbol{A}^{\mathrm{H}}. \end{aligned} \tag{A.21}$$

A.2.4 Fourth-Order Moment

In this subsection, we compute the moment $\mathsf{E}\left\{\|\boldsymbol{z}\|^4\right\}$, which is used in many places in the text.

Consider first the scalar case, and let $z \sim \mathrm{CN}(0, 1)$. Then

$$\begin{aligned}\mathsf{E}\left\{|z|^4\right\} &= \frac{1}{\pi}\int_{\mathbb{C}} dz\ |z|^4 e^{-|z|^2}\\ &= \frac{1}{\pi}\int_0^{\infty}\int_{-\pi}^{\pi} dr\ d\theta\ r^5 e^{-r^2}\\ &= 2\int_0^{\infty} dr\ r^5 e^{-r^2}\\ &= \int_0^{\infty} du\ u^2 e^{-u}\\ &= 2. \end{aligned} \tag{A.22}$$

Next, in the vector-valued case, let $\boldsymbol{z} = [z_1, \ldots, z_M]^T$ be a vector with distribution $\boldsymbol{z} \sim \mathrm{CN}(\boldsymbol{0}, \boldsymbol{I}_M)$. Then

$$\begin{aligned}\mathsf{E}\left\{\|\boldsymbol{z}\|^4\right\} &= \mathsf{E}\left\{\left(\sum_{m=1}^{M} |z_m|^2\right)^2\right\}\\ &= \sum_{m=1}^{M}\sum_{m'=1}^{M} \mathsf{E}\left\{|z_m|^2|z_{m'}|^2\right\}\\ &= \sum_{m=1}^{M} \mathsf{E}\left\{|z_m|^4\right\} + \sum_{m=1}^{M}\sum_{\substack{m'=1\\ m'\neq m}}^{M} \mathsf{E}\left\{|z_m|^2|z_{m'}|^2\right\}\\ &= M(M+1), \end{aligned} \tag{A.23}$$

where in the last equality we used (A.22).

Appendix B

USEFUL RANDOM MATRIX RESULTS

Let $\boldsymbol{Z}$ be an $M \times K$, $M \geq K$, matrix whose elements are i.i.d. CN(0, 1) random variables. In this appendix, we provide self-contained proofs of the following two facts:

$$\mathsf{E}\left\{\left[\left(\boldsymbol{Z}^{\mathrm{H}}\boldsymbol{Z}\right)^{-1}\right]_{kk}\right\} = \frac{1}{M-K}, \quad k = 1, \ldots, K, \tag{B.1}$$

$$\mathsf{E}\left\{\frac{1}{\left[\left(\boldsymbol{Z}^{\mathrm{H}}\boldsymbol{Z}\right)^{-1}\right]_{kk}}\right\} = M + 1 - K, \quad k = 1, \ldots, K. \tag{B.2}$$

Equations (B.1)–(B.2) are the two principal random matrix-theoretic results used in this book. Strictly speaking, there is no new material in this appendix, but we are not aware of a simpler treatment elsewhere.

We first show in Section B.1 that all diagonal elements of $\left(\boldsymbol{Z}^{\mathrm{H}}\boldsymbol{Z}\right)^{-1}$ have the same marginal distributions. In Section B.2, we then obtain the explicit marginal density. Finally, in Section B.3 we compute the expectations in (B.1)–(B.2).

B.1 Symmetry

Let $\boldsymbol{P}$ be a $K \times K$ permutation matrix – a unitary matrix that shuffles the order of the elements of a vector. Pre-multiplication and post-multiplication of a $K \times K$ matrix by $\boldsymbol{P}$ and $\boldsymbol{P}^{\mathrm{H}}$ respectively shuffles the elements of the matrix while retaining the same diagonal elements in shuffled form. We shuffle the inverse of the Gramian of $\boldsymbol{Z}$ as follows:

$$\begin{aligned}\boldsymbol{P}\left(\boldsymbol{Z}^{\mathrm{H}}\boldsymbol{Z}\right)^{-1}\boldsymbol{P}^{\mathrm{H}} &= \left(\boldsymbol{P}\boldsymbol{Z}^{\mathrm{H}}\boldsymbol{Z}\boldsymbol{P}^{\mathrm{H}}\right)^{-1} \\ &= \left(\boldsymbol{Z}'^{\mathrm{H}}\boldsymbol{Z}'\right)^{-1},\end{aligned} \tag{B.3}$$

where $\boldsymbol{Z}' = \boldsymbol{Z}\boldsymbol{P}^{\mathrm{H}}$. But the density of $\boldsymbol{Z}$ is invariant to multiplications by a unitary matrix, from which we conclude that the distribution of the shuffled matrix $\boldsymbol{P}\left(\boldsymbol{Z}^{\mathrm{H}}\boldsymbol{Z}\right)^{-1}\boldsymbol{P}^{\mathrm{H}}$ is the same as that of the original matrix $\left(\boldsymbol{Z}^{\mathrm{H}}\boldsymbol{Z}\right)^{-1}$. Hence the diagonal elements of $\left(\boldsymbol{Z}^{\mathrm{H}}\boldsymbol{Z}\right)^{-1}$ have identical marginal distributions.

B.2 QR Factorization

The QR factorization represents the matrix $\boldsymbol{Z}$ as the product of an $M \times M$ unitary matrix $\boldsymbol{Q}$ and an $M \times K$ upper-triangular matrix $\boldsymbol{R}$,

$$\begin{aligned} \boldsymbol{Z} &= \boldsymbol{Q}\boldsymbol{R} \\ &= \boldsymbol{Q}\begin{bmatrix} \bar{\boldsymbol{R}} \\ \boldsymbol{0} \end{bmatrix}, \end{aligned} \tag{B.4}$$

where $\bar{\boldsymbol{R}}$ is $K \times K$ upper-triangular. In terms of this factorization, we have

$$\begin{aligned} \left(\boldsymbol{Z}^{\mathrm{H}}\boldsymbol{Z}\right)^{-1} &= \left(\boldsymbol{R}^{\mathrm{H}}\boldsymbol{Q}^{\mathrm{H}}\boldsymbol{Q}\boldsymbol{R}\right)^{-1} \\ &= \left(\boldsymbol{R}^{\mathrm{H}}\boldsymbol{R}\right)^{-1} \\ &= \left(\bar{\boldsymbol{R}}^{\mathrm{H}}\bar{\boldsymbol{R}}\right)^{-1} \\ &= \bar{\boldsymbol{R}}^{-1}\left(\bar{\boldsymbol{R}}^{-1}\right)^{\mathrm{H}}. \end{aligned} \tag{B.5}$$

We focus on the Kth diagonal element, and recall that the inverse of an upper-triangular matrix is also upper-triangular,

$$\begin{aligned} \left[\left(\boldsymbol{Z}^{\mathrm{H}}\boldsymbol{Z}\right)^{-1}\right]_{KK} &= \left[\bar{\boldsymbol{R}}^{-1}\left(\bar{\boldsymbol{R}}^{-1}\right)^{\mathrm{H}}\right]_{KK} \\ &= \left[\bar{\boldsymbol{R}}^{-1}\right]_{KK}\left[\left(\bar{\boldsymbol{R}}^{-1}\right)^{\mathrm{H}}\right]_{KK} \\ &= \frac{1}{|[\bar{\boldsymbol{R}}]_{KK}|^2}, \end{aligned} \tag{B.6}$$

where $[\bar{\boldsymbol{R}}]_{KK}$ is the (K, K)th element of $\bar{\boldsymbol{R}}$.

We construct the unitary factor, $\boldsymbol{Q}$, as a product of complex Householder matrices. Given an $M \times 1$ unitary vector $\boldsymbol{a}$, $\|\boldsymbol{a}\|^2 = 1$, the following unitary matrix:

$$\boldsymbol{H} = \boldsymbol{I}_M - \frac{(\boldsymbol{e}_1 - \boldsymbol{a})(\boldsymbol{e}_1 - \boldsymbol{a})^{\mathrm{H}}}{1 - \boldsymbol{a}^{\mathrm{H}}\boldsymbol{e}_1}, \tag{B.7}$$

collapses the vector $\boldsymbol{a}$ into its first entry,

$$\boldsymbol{H}^{\mathrm{H}}\boldsymbol{a} = \boldsymbol{e}_1, \tag{B.8}$$

where $\boldsymbol{e}_1$ is the first column of $\boldsymbol{I}_M$.

We form a unitary vector from the first column of the matrix $\boldsymbol{Z}$, $\boldsymbol{a} = z_1/\|z_1\|$, and construct the associated Householder matrix, denoted $\boldsymbol{H}_1$. The multiplication of the matrix $\boldsymbol{Z}$ by $\boldsymbol{H}_1^{\mathrm{H}}$ from the left does two things: first, it collapses the first column into its first entry, $\boldsymbol{H}_1^{\mathrm{H}} z_1 = \|z_1\|\, \boldsymbol{e}_1$; second, it transforms the remaining columns of $\boldsymbol{Z}$ without altering their statistical properties. Specifically, conditioned on z_1, the product of $\boldsymbol{H}_1^{\mathrm{H}}$ and the remaining columns of $\boldsymbol{Z}$ comprise independent $\mathrm{CN}(0, 1)$ elements because the distribution of independent $\mathrm{CN}(0, 1)$ random variables is invariant to unitary transformations. This conditional density does not depend on $\boldsymbol{H}_1$ nor does it depend on z_1, and hence the entities are statistically independent. Note that $\|z_1\|^2$ is the sum of absolute squares of M $\mathrm{CN}(0, 1)$ random variables, so it is equal to one-half of a chi-square random variable with $2M$ degrees of freedom. Symbolically, we summarize the action of the Householder matrix as follows:

$$\boldsymbol{H}_1^{\mathrm{H}}\boldsymbol{Z} = \begin{bmatrix} \sqrt{\frac{1}{2}\chi^2_{2M}} & \mathrm{CN}(0,1) & \cdots & \mathrm{CN}(0,1) \\ 0 & \mathrm{CN}(0,1) & \cdots & \mathrm{CN}(0,1) \\ \vdots & \vdots & \cdots & \vdots \\ 0 & \mathrm{CN}(0,1) & \cdots & \mathrm{CN}(0,1) \end{bmatrix}, \tag{B.9}$$

where it is understood that all elements of the matrix are statistically independent. Next we apply a Householder matrix that leaves the first column and first row of $\boldsymbol{H}_1^{\mathrm{H}}\boldsymbol{Z}$ untouched, but which collapses the last $M - 1$ elements of the second column,

$$\boldsymbol{H}_2 = \begin{bmatrix} 1 & \boldsymbol{0}^{\mathrm{H}} \\ \boldsymbol{0} & \bar{\boldsymbol{H}}_2 \end{bmatrix}, \tag{B.10}$$

where $\bar{\boldsymbol{H}}_2$ is an $(M-1) \times (M-1)$ Householder matrix. The application of this second Householder transformation yields the following:

$$\boldsymbol{H}_2^{\mathrm{H}}\boldsymbol{H}_1^{\mathrm{H}}\boldsymbol{Z} = \begin{bmatrix} \sqrt{\frac{1}{2}\chi^2_{2M}} & \mathrm{CN}(0,1) & \mathrm{CN}(0,1) & \cdots & \mathrm{CN}(0,1) \\ 0 & \sqrt{\frac{1}{2}\chi^2_{2(M-1)}} & \mathrm{CN}(0,1) & \cdots & \mathrm{CN}(0,1) \\ 0 & 0 & \mathrm{CN}(0,1) & \cdots & \mathrm{CN}(0,1) \\ \vdots & \vdots & \vdots & \cdots & \vdots \\ 0 & 0 & \mathrm{CN}(0,1) & \cdots & \mathrm{CN}(0,1) \end{bmatrix}. \tag{B.11}$$

We proceed in analogous fashion, and after the application of the Kth and final Householder transformation we have

$$\boldsymbol{H}_K^{\mathrm{H}} \cdots \boldsymbol{H}_2^{\mathrm{H}} \boldsymbol{H}_1^{\mathrm{H}} \mathbf{Z} = \begin{bmatrix} \sqrt{\frac{1}{2}\chi^2_{2M}} & \mathrm{CN}(0,1) & \cdots & \mathrm{CN}(0,1) \\ 0 & \sqrt{\frac{1}{2}\chi^2_{2(M-1)}} & \cdots & \mathrm{CN}(0,1) \\ \vdots & \vdots & \ddots & \sqrt{\frac{1}{2}\chi^2_{2(M+1-K)}} \\ 0 & 0 & \cdots & 0 \\ \vdots & \vdots & \cdots & \vdots \\ 0 & 0 & \cdots & 0 \end{bmatrix}. \tag{B.12}$$

Hence we have shown that $|[\bar{\boldsymbol{R}}]_{KK}|^2$ has the same distribution as the sum of absolute squares of $M+1-K$ independent $\mathrm{CN}(0,1)$ random variables.

B.3 Expectations

In what follows, let $\boldsymbol{y}$ denote an $(M+1-K)\times 1$ vector of independent $\mathrm{CN}(0,1)$ random variables. Given (B.6), the symmetry properties of $\left(\mathbf{Z}^{\mathrm{H}}\mathbf{Z}\right)^{-1}$ and the distribution derived in Section B.2, we can now compute the first desired expectation. We obtain, for $k=1,\ldots,K$,

$$\begin{aligned} \mathsf{E}\left\{\left[\left(\mathbf{Z}^{\mathrm{H}}\mathbf{Z}\right)^{-1}\right]_{kk}\right\} &= \mathsf{E}\left\{\left[\left(\mathbf{Z}^{\mathrm{H}}\mathbf{Z}\right)^{-1}\right]_{KK}\right\} \\ &= \mathsf{E}\left\{\frac{1}{|[\bar{\boldsymbol{R}}]_{KK}|^2}\right\} \\ &= \mathsf{E}\left\{\frac{1}{\|\boldsymbol{y}\|^2}\right\} \\ &= \frac{1}{\pi^{M+1-K}}\int_{\mathbb{C}^{M+1-K}} d\boldsymbol{y}\ \frac{\exp\left(-\|\boldsymbol{y}\|^2\right)}{\|\boldsymbol{y}\|^2}. \end{aligned} \tag{B.13}$$

At this point, we utilize the identity,

$$\frac{1}{\|\boldsymbol{y}\|^2} = \int_0^\infty ds\ \exp\left(-s\|\boldsymbol{y}\|^2\right) \tag{B.14}$$

within (B.13). After changing orders of integration,[1] we obtain the final result,

$$\begin{aligned}\mathsf{E}\left\{\left[\left(\mathbf{Z}^{\mathrm{H}}\mathbf{Z}\right)^{-1}\right]_{kk}\right\} &= \int_0^\infty ds \, \frac{1}{\pi^{M+1-K}} \int_{\mathbb{C}^{M+1-K}} d\boldsymbol{y} \, \exp\left(-s\|\boldsymbol{y}\|^2\right) \exp\left(-\|\boldsymbol{y}\|^2\right) \\ &= \int_0^\infty ds \, \frac{1}{(1+s)^{M+1-K}} \\ &= \frac{1}{M-K}, \quad k = 1, \ldots, K. \end{aligned} \tag{B.15}$$

The second required expectation (that of the double inverse) is simpler,

$$\begin{aligned}\mathsf{E}\left\{\frac{1}{\left[\left(\mathbf{Z}^{\mathrm{H}}\mathbf{Z}\right)^{-1}\right]_{kk}}\right\} &= \mathsf{E}\left\{\frac{1}{\left[\left(\mathbf{Z}^{\mathrm{H}}\mathbf{Z}\right)^{-1}\right]_{KK}}\right\} \\ &= \mathsf{E}\left\{|[\bar{\boldsymbol{R}}]_{KK}|^2\right\} \\ &= M+1-K, \quad k = 1, \ldots, K. \end{aligned} \tag{B.16}$$

[1]Interchanging the order of integration is possible here since the integrand is positive and continuous.

Appendix C

CAPACITY AND CAPACITY BOUNDING TOOLS

This appendix reviews some basic results related to capacity and capacity bounds for scalar point-to-point, Point-to-Point MIMO and Multiuser MIMO channels.

C.1 Jensen's Inequality

Jensen's inequality [146, Theorem 2.6.2] is a useful formula which states that for any random variable z, and an arbitrary convex $\cap$ function $f(\cdot)$ (see Figure C.1),

$$\mathsf{E}\{f(z)\} \leq f(\mathsf{E}\{z\}). \tag{C.1}$$

Two instances of (C.1) are particularly useful:

- With $f(z) = \log_2(1+z)$, for $z > 0$, the second derivative of $\log_2(1+z)$ with respect to z is negative, so $f(z)$ is convex $\cap$. Equation (C.1) gives directly

$$\mathsf{E}\{\log_2(1+z)\} \leq \log_2(1+\mathsf{E}\{z\}). \tag{C.2}$$

- With $z = 1/u$ for $z > 0$, the second derivative of the function $\log_2(1+1/u)$ with respect to u is positive. Consequently, $-\log_2(1+1/u)$ is convex $\cap$. The application of (C.1) gives

$$\mathsf{E}\{\log_2(1+z)\} \geq \log_2\left(1+\frac{1}{\mathsf{E}\left\{\frac{1}{z}\right\}}\right). \tag{C.3}$$

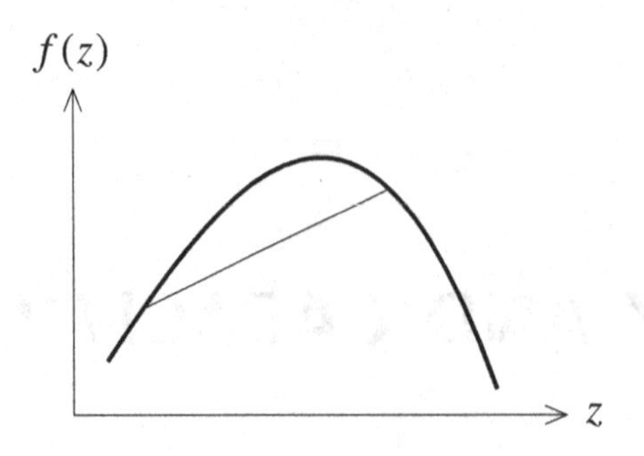

Figure C.1. A function $f(\cdot)$ is convex $\cap$ if any chord lies below the curve.

C.2 Point-to-Point Scalar Channel

This section complements Section 2.3, where point-to-point scalar channels were introduced. Figure 2.9 recapitulates the different channel models of concern, and Table 2.3 summarizes the main results.

The capacity of a general point-to-point scalar channel is [146, 147][1]

$$C = \max_{\substack{p_x(\cdot) \\ \mathsf{E}\{|x|^2\} \leq 1}} \mathsf{I}\{y; x\}, \tag{C.4}$$

where $\mathsf{E}\left\{|x|^2\right\} \leq 1$ represents a power constraint[2] and $\mathsf{I}\{y; x\}$ is a quantity called the *mutual information* between the channel input x and its output y,

$$\mathsf{I}\{y; x\} = \mathsf{h}\{y\} - \mathsf{h}\{y|x\}, \tag{C.5}$$

where $\mathsf{h}\{y\}$ and $\mathsf{h}\{y|x\}$ are defined as follows:

$$\begin{aligned} \mathsf{h}\{y\} &= -\mathsf{E}\left\{\log_2\left(p_y(y)\right)\right\} \\ &= -\int dt\, p_y(t) \log_2\left(p_y(t)\right), \end{aligned} \tag{C.6}$$

$$\begin{aligned} \mathsf{h}\{y|x\} &= -\mathsf{E}\left\{\log_2\left(p_{y|x}(y|x)\right)\right\} \\ &= -\int dt'\, p_x(t') \int dt\, p_{y|x}(t|t') \log_2\left(p_{y|x}(t|t')\right). \end{aligned} \tag{C.7}$$

[1]The notation $p_x(\cdot)$ and the term "probability density" imply that x is a continuous random variable. However, all random variables of concern in this chapter could have discrete or mixed distributions, in which case similar definitions and results apply but additional notation is needed for a mathematically rigorous treatment.

[2]The power normalization is arbitrary – as in Section 2.1.8 we can assume a unit power constraint and adjust the SNR by scaling the transmitted symbol by a factor $\sqrt{\rho}$.

The quantities $\mathsf{h}\{y\}$ and $\mathsf{h}\{y|x\}$ are called *differential entropy* and *conditional differential entropy*, respectively. In (C.4), the maximization is with respect to the probability distribution $p_x(\cdot)$, subject to the power constraint. One can show that $\mathsf{I}\{y;x\}$ is symmetric, in the sense that

$$\begin{aligned}\mathsf{I}\{y;x\} &= \mathsf{I}\{x;y\} \\ &= \mathsf{h}\{x\} - \mathsf{h}\{x|y\}. \end{aligned} \tag{C.8}$$

For vector-valued random variables $\boldsymbol{x}$ and $\boldsymbol{y}$, $\mathsf{I}\{\boldsymbol{y};\boldsymbol{x}\}$, $\mathsf{h}\{\boldsymbol{x}\}$ and $\mathsf{h}\{\boldsymbol{y}|\boldsymbol{x}\}$ are defined similarly.

Although the quantity $\mathsf{I}\{x;y\}$ has some intuitive appeal – the reduction in uncertainty (entropy) of x due to the measurement of y – it only takes on concrete meaning in the context of one of two coding theorems. The *noisy channel coding theorem* concerns the possibility of reliable transmission at any rate less than capacity, while the *converse to the coding theorem* asserts the impossibility of reliable transmission at rates greater than capacity. Proofs of the noisy channel coding theorem entail a random codebook which maps sequences of message bits into sequences of statistically independent symbols chosen according to the probability density $p_x(\cdot)$.

C.2.1 Entropy is Invariant to Translation

The entropy of a random variable x is invariant to translation in the sense that $\mathsf{h}\{x+a\} = \mathsf{h}\{x\}$ for any constant a. To see this, note that if $x' = x + a$, then $p_{x'}(t) = p_x(t-a)$. Hence,

$$\begin{aligned}\mathsf{h}\{x'\} &= -\int dt\; p_{x'}(t) \log_2\left(p_{x'}(t)\right) \\ &= -\int d(t-a)\; p_x(t-a) \log_2\left(p_x(t-a)\right) \\ &= -\int dt\; p_x(t) \log_2\left(p_x(t)\right) \\ &= \mathsf{h}\{x\}. \end{aligned} \tag{C.9}$$

Similarly, conditional entropy is invariant to translation in the sense that $\mathsf{h}\{x+a+by|y\} = \mathsf{h}\{x|y\}$ for any two random variables x and y and constants a and b.

C.2.2 The Gaussian Distribution Maximizes Entropy

We first derive a subsidiary result that will be used repeatedly. Let x be a random variable that satisfies $\mathsf{Var}\{x\} \leq 1$. Then $\mathsf{h}\{x\}$ is maximized if $x \sim \mathrm{CN}(0,1)$. To establish this fact, first note that by the translation invariance property (Section C.2.1), the mean of x does not affect $\mathsf{h}\{x\}$. Hence, we can without loss of generality assume that $\mathsf{E}\{x\} = 0$, in which case the constraint $\mathsf{Var}\{x\} \leq 1$ is equivalent to $\mathsf{E}\left\{|x|^2\right\} \leq 1$. Next define the auxiliary function $q(t)$ according to

$$q(t) = \frac{1}{\pi} e^{-|t|^2}. \tag{C.10}$$

Note that $q(t)$ is formally identical to the probability density function of a $\mathrm{CN}(0, 1)$ random variable. Then upper bound $\mathsf{h}\{x\}$ as follows:

$$\begin{aligned}
\mathsf{h}\{x\} &= -\mathsf{E}\{\log_2(p_x(x))\} \\
&= \mathsf{E}\left\{\log_2\left(\frac{q(x)}{p_x(x)}\right)\right\} - \mathsf{E}\{\log_2(q(x))\} \\
&\overset{(a)}{\leq} \log_2\left(\mathsf{E}\left\{\frac{q(x)}{p_x(x)}\right\}\right) - \mathsf{E}\{\log_2(q(x))\} \\
&\overset{(b)}{=} 0 - \mathsf{E}\{\log_2(q(x))\} \\
&\overset{(c)}{=} -\mathsf{E}\left\{-\log_2(\pi) - |x|^2 \log_2(e)\right\} \\
&\overset{(d)}{\leq} \log_2(\pi e),
\end{aligned} \tag{C.11}$$

where the expectations are taken with respect to the probability density function $p_x(\cdot)$. In (C.11), (a) follows from Jensen's inequality, (C.1), together with the fact that the logarithm is convex $\cap$, (b) follows from the fact that

$$\begin{aligned}
\mathsf{E}\left\{\frac{q(x)}{p_x(x)}\right\} &= \int dt\, p_x(t)\frac{q(t)}{p_x(t)} \\
&= 1,
\end{aligned} \tag{C.12}$$

in (c) we inserted (C.10), and in (d) we used the constraint $\mathsf{E}\left\{|x|^2\right\} \leq 1$. Equality in (a)–(d) holds precisely if $p_x(t) = q(t)$, in which case $x \sim \mathrm{CN}(0, 1)$.

More generally, for any random variable x with given variance, an upper bound on entropy is $\mathsf{h}\{x\} \leq \log_2(\pi e \mathsf{Var}\{x\})$ with equality if x is circularly symmetric Gaussian with zero mean. Also, if x and y are two, possibly statistically dependent, random variables, then the conditional entropy satisfies

$$\mathsf{h}\{x|y\} \leq \mathsf{E}\{\log_2(\pi e \mathsf{Var}\{x|y\})\}, \tag{C.13}$$

with equality if the conditional density $p_{x|y}(\cdot|\cdot)$ is circularly symmetric Gaussian.

In the vector-valued case, for given $\mathsf{Cov}\left\{\boldsymbol{x}\right\}$,

$$\mathsf{h}\left\{\boldsymbol{x}\right\} \leq \log_2\left|\pi e \mathsf{Cov}\left\{\boldsymbol{x}\boldsymbol{x}^H\right\}\right|, \quad \text{(C.14)}$$

with equality if $\boldsymbol{x}$ is circularly symmetric Gaussian. The above proof applies in this case as well, where the function $q(\cdot)$ is the probability density function of a multivariate circularly symmetric Gaussian vector having the specified covariance.

C.2.3 Deterministic Channel with Additive Gaussian Noise

Consider the deterministic channel with additive Gaussian noise; see Section 2.3.1 and Figure 2.9(a). To perform the maximization in (C.4), first note that the constraint $\mathsf{E}\left\{|x|^2\right\} \leq 1$ implies that $\mathsf{E}\left\{|y|^2\right\} \leq 1+\rho$. Then write $\mathsf{I}\left\{y;x\right\}$ as follows:

$$\begin{aligned}\mathsf{I}\left\{y;x\right\} &= \mathsf{h}\left\{y\right\} - \mathsf{h}\left\{y|x\right\} \\ &\overset{(a)}{=} \mathsf{h}\left\{y\right\} - \mathsf{h}\left\{y-\sqrt{\rho}x|x\right\} \\ &\overset{(b)}{=} \mathsf{h}\left\{y\right\} - \mathsf{h}\left\{w\right\} \\ &= \mathsf{h}\left\{y\right\} - \log_2(\pi e) \\ &\overset{(c)}{\leq} \log_2\left(1+\rho\right),\end{aligned} \quad \text{(C.15)}$$

where in (a) we used the fact that translation does not change the entropy (see Section C.2.1), (b) follows from the fact that x and w are independent so $\mathsf{h}\left\{y-\sqrt{\rho}x|x\right\} = \mathsf{h}\left\{w|x\right\} = \mathsf{h}\left\{w\right\}$, and (c) follows by applying the result obtained in Section C.2.2. Equality in (c) holds precisely if $y \sim \mathrm{CN}(0, 1+\rho)$. The particular choice of input distribution $x \sim \mathrm{CN}(0,1)$ satisfies the constraint $\mathsf{E}\left\{|x|^2\right\} \leq 1$ and yields $y \sim \mathrm{CN}(0, 1+\rho)$. Hence, under the given constraints the maximum possible value of $\mathsf{I}\left\{y;x\right\}$ is C in (2.38), and the distribution of x that achieves this capacity is $x \sim \mathrm{CN}(0,1)$.

C.2.4 Deterministic Channel with Additive Non-Gaussian Noise

The next case of concern is that of a deterministic channel with additive non-Gaussian noise; see Section 2.3.2 and Figure 2.9(b). In what follows, we derive the capacity bound (2.40), which is substantially a special case of results in [148, 149].

First, note that for any distribution $p_x(\cdot)$,

$$\begin{aligned} C &\geq \mathsf{I}\{y;x\} \\ &= \mathsf{h}\{x\} - \mathsf{h}\{x|y\} \\ &\overset{(a)}{\geq} \mathsf{h}\{x\} - \mathsf{E}\{\log_2(\pi e \mathsf{Var}\{x|y\})\} \\ &\overset{(b)}{\geq} \mathsf{h}\{x\} - \log_2(\pi e \mathsf{E}\{\mathsf{Var}\{x|y\}\}) \\ &= \mathsf{h}\{x\} - \log_2\left(\pi e \mathsf{E}\left\{|x - \mathsf{E}\{x|y\}|^2\right\}\right) \\ &\overset{(c)}{\geq} \mathsf{h}\{x\} - \log_2\left(\pi e \frac{\mathsf{Var}\{x\}}{1 + \rho \mathsf{Var}\{x\}}\right), \end{aligned} \tag{C.16}$$

where in (a) we used the result in Section C.2.2, in (b) we used Jensen's inequality (see (C.2)), in (c) we used the fact that $\mathsf{E}\left\{|x - \mathsf{E}\{x|y\}|^2\right\}$ is the mean-square error associated with the MMSE estimate of x given y, which cannot exceed that of the *linear* MMSE estimator, which in turn is equal to $\mathsf{Var}\{x\}/(1 + \rho\mathsf{Var}\{x\})$. In (c), the fact that x and w are uncorrelated is crucial.

The bound in (C.16) holds irrespective of the distribution $p_x(\cdot)$. By taking x to be $\mathrm{CN}(0, 1)$, the power constraint $\mathsf{E}\left\{|x|^2\right\} \leq 1$ is satisfied and we have $\mathsf{h}\{x\} = \log_2(\pi e)$ and $\mathsf{Var}\{x\} = 1$. With this choice of $p_x(\cdot)$, (C.16) yields

$$\begin{aligned} C &\geq \log_2(\pi e) - \log_2\left(\pi e \frac{1}{1+\rho}\right) \\ &= \log_2(1 + \rho), \end{aligned} \tag{C.17}$$

which is (2.40). We stress, however, that the optimal input distribution $p_x(\cdot)$ is in general not Gaussian.

The above capacity bound is quite powerful, and is sufficient to derive all of the principal capacity bounds in Chapters 3 and 4. On the downlink, Bayesian statistics represent the received signal as a known gain times the desired signal, plus uncorrelated noise. On the uplink, the base station first performs linear decoding, then it conveniently forgets that it knows the channel estimates and uses Bayesian statistics to represent the processed signal for each terminal as a known gain times the desired signal, plus uncorrelated noise.

C.2.5 Fading Channel with Additive Gaussian Noise and Perfect CSI at the Receiver

Next, consider the point-to-point, scalar fading channel with additive Gaussian noise defined in Section 2.3.3 and Figure 2.9(c). The receiver knows g, and hence effectively observes

(y, g) instead of only y. Hence, the pertinent mutual information in this case – to be used in lieu of $\mathsf{I}\{y; x\}$ – is $\mathsf{I}\{y, g; x\}$. We calculate $\mathsf{I}\{y, g; x\}$ as follows:

$$\begin{aligned}
\mathsf{I}\{y, g; x\} &= \mathsf{h}\{y, g\} - \mathsf{h}\{y, g|x\} \\
&= \mathsf{E}\left\{\log_2\left(\frac{p_{y,g|x}(y, g|x)}{p_{y,g}(y, g)}\right)\right\} \\
&= \mathsf{E}\left\{\log_2\left(\frac{p_{y|g,x}(y|g, x)p_{g,x}(g, x)}{p_x(x)p_{y,g}(y, g)}\right)\right\} \\
&\overset{(a)}{=} \mathsf{E}\left\{\log_2\left(\frac{p_{y|g,x}(y|g, x)p_g(g)p_x(x)}{p_x(x)p_{y,g}(y, g)}\right)\right\} \\
&= \mathsf{E}\left\{\log_2\left(\frac{p_{y|g,x}(y|g, x)}{p_{y|g}(y|g)}\right)\right\} \\
&= \mathsf{E}\left\{\mathsf{E}\left\{\log_2\left(\frac{p_{y|g,x}(y|g, x)}{p_{y|g}(y|g)}\right)\middle| g\right\}\right\},
\end{aligned} \tag{C.18}$$

where in (a) we exploited the fact that x and g are independent. The outer expectation on the right-hand side of (C.18) is with respect to g. In view of the results in Section C.2.3, the expression inside the outer expectation on the right-hand side of (C.18) is maximized for any g by taking $x \sim \mathcal{CN}(0, 1)$, and (2.42) follows.

C.2.6 Fading Channel with Additive Non-Gaussian Noise and Side Information

The final case of concern is that of a fading channel with non-Gaussian additive noise and side information; see Section 2.3.5 and Figure 2.9(e).

The relevant mutual information between x and (y, Ω) is $\mathsf{I}\{y, \Omega; x\}$. The optimal input distribution $p_x(\cdot)$, and the resulting capacity C, are unknown in general. To derive a bound, note that for any distribution of x,

$$\begin{aligned}
C &\geq \mathsf{I}\{y, \Omega; x\} \\
&= \mathsf{h}\{x\} - \mathsf{h}\{x|y, \Omega\} \\
&= \mathsf{E}\left\{\log_2\left(\frac{p_{x|y,\Omega}(x|y, \Omega)}{p_x(x)}\right)\right\} \\
&\overset{(a)}{=} \mathsf{E}\left\{\log_2\left(\frac{p_{x|y,\Omega}(x|y, \Omega)}{p_{x|\Omega}(x|\Omega)}\right)\right\} \\
&= \mathsf{E}\left\{\mathsf{E}\left\{\log_2\left(\frac{p_{x|y,\Omega}(x|y, \Omega)}{p_{x|\Omega}(x|\Omega)}\right)\middle| \Omega\right\}\right\},
\end{aligned} \tag{C.19}$$

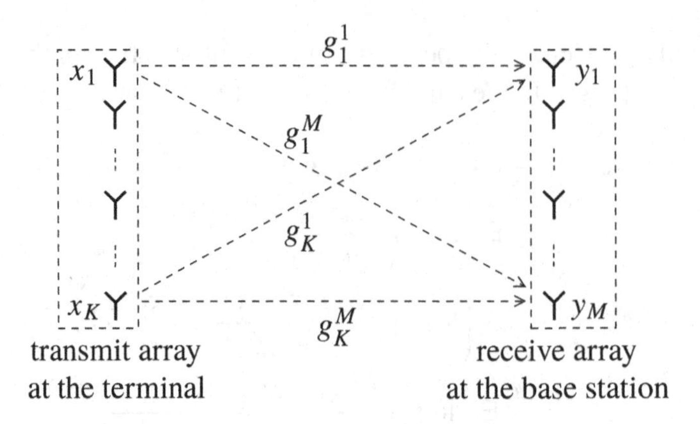

Figure C.2. Uplink Point-to-Point MIMO channel.

where in (a) we used that x and Ω are independent. The right-hand side of (C.19) comprises an outer expectation with respect to Ω; everything inside is as in Section C.2.4. Hence, for every value of Ω we can lower-bound the conditional expectation using the chain of inequalities in (C.16). This yields, irrespective of the distribution of x,

$$\begin{aligned}
&\mathsf{E}\left\{\log_2\left(\frac{p_{x|y,\Omega}(x|y,\Omega)}{p_{x|\Omega}(x|\Omega)}\right)\middle|\Omega\right\} \\
&\geq \mathsf{h}\{x\} - \log_2\left(\pi e \mathsf{E}\{\mathsf{Var}\{x|y,\Omega\}|\,\Omega\}\right) \\
&\geq \mathsf{h}\{x\} - \log_2\left(\pi e \mathsf{E}\left\{|x - \mathsf{E}\{x|y,\Omega\}|^2\middle|\Omega\right\}\right) \\
&\overset{(a)}{\geq} \mathsf{h}\{x\} - \log_2\left(\pi e \mathsf{Var}\{x\} - \pi e \frac{\rho\,|\mathsf{E}\{g|\Omega\}|^2\,(\mathsf{Var}\{x\})^2}{\rho\,|\mathsf{E}\{g|\Omega\}|^2\,\mathsf{Var}\{x\} + \rho \mathsf{Var}\{g|\Omega\}\mathsf{E}\{|x|^2\} + \mathsf{Var}\{w|\Omega\}}\right),
\end{aligned} \tag{C.20}$$

where in (a) we used (2.45), and when computing the mean-square error of the linear MMSE estimate conditioned on Ω, we wrote y in (2.41) in terms of a term comprising a deterministic gain multiplied by x plus two mutually uncorrelated terms:

$$y = \sqrt{\rho}\mathsf{E}\{g|\Omega\}x + \sqrt{\rho}(g - \mathsf{E}\{g|\Omega\})x + w. \tag{C.21}$$

By taking x to be CN(0, 1), the constraint $\mathsf{E}\left\{|x|^2\right\} \leq 1$ is satisfied and $\mathsf{h}\{x\} = \log_2(\pi e)$. The expectation of the right-hand side of (C.20) with respect to Ω and its subsequent simplification gives (2.46).

C.3 Point-to-Point MIMO Channel

With reference to Section 1.1, the Point-to-Point MIMO channel arises when both the base station and the terminal use arrays of phase-coherent cooperating antennas. For the sake of argument, consider the uplink, that is, the transmission from a terminal with K antennas to a base station with M antennas; see Figure C.2. The signal model is formally identical to that of uplink Multi-User MIMO (Section 2.2.1):

$$\boldsymbol{y} = \sqrt{\rho}\boldsymbol{G}\boldsymbol{x} + \boldsymbol{w}, \tag{C.22}$$

with the crucial difference that now a central entity creates the vector of transmitted signals: $\boldsymbol{y} = [y_1, \dots, y_M]^{\mathrm{T}}$ is a vector of receive signals, $\boldsymbol{x} \triangleq [x_1, \dots, x_K]^{\mathrm{T}}$ is a vector of transmit signals, $\boldsymbol{w} = [w_1, \dots, w_M]^{\mathrm{T}}$ comprises i.i.d. CN(0, 1) noise, and

$$\boldsymbol{G} = \begin{bmatrix} g_1^1 & \cdots & g_K^1 \\ \vdots & \ddots & \vdots \\ g_1^M & \cdots & g_K^M \end{bmatrix}, \tag{C.23}$$

where g_k^m is the channel gain between terminal antenna k and base station antenna m.

C.3.1 Deterministic MIMO Channel

We first consider the case that $\boldsymbol{G}$ is a deterministic constant, and hence is known to both the terminal and the base station. The capacity is obtained by maximizing the mutual information between $\boldsymbol{x}$ and $\boldsymbol{y}$ with respect to $p_{\boldsymbol{x}}(\cdot)$, under a constraint on the total emitted power,

$$\mathsf{E}\left\{\|\boldsymbol{x}\|^2\right\} \leq 1. \tag{C.24}$$

Similarly to Section C.2.3, for any $p_{\boldsymbol{x}}(\cdot)$,

$$\begin{aligned}
\mathsf{I}\{\boldsymbol{y};\boldsymbol{x}\} &= \mathsf{h}\{\boldsymbol{y}\} - \mathsf{h}\{\boldsymbol{y}|\boldsymbol{x}\} \\
&= \mathsf{h}\{\boldsymbol{y}\} - \mathsf{h}\{\boldsymbol{y} - \sqrt{\rho}\boldsymbol{G}\boldsymbol{x}|\boldsymbol{x}\} \\
&= \mathsf{h}\{\boldsymbol{y}\} - \mathsf{h}\{\boldsymbol{w}\} \\
&= \mathsf{h}\{\boldsymbol{y}\} - \log_2|\pi e \boldsymbol{I}_M| \\
&\overset{(a)}{\leq} \log_2\left|\pi e\left(\boldsymbol{I}_M + \rho\boldsymbol{G}\boldsymbol{Q}\boldsymbol{G}^{\mathrm{H}}\right)\right| - \log_2|\pi e\boldsymbol{I}_M| \\
&= \log_2\left|\boldsymbol{I}_M + \rho\boldsymbol{G}\boldsymbol{Q}\boldsymbol{G}^{\mathrm{H}}\right|,
\end{aligned} \tag{C.25}$$

where we defined

$$Q = \mathsf{Cov}\{x\}, \tag{C.26}$$

and where in (a) we have equality if y is circularly symmetric Gaussian – see Section C.2.2 – which is the case if x is circularly symmetric Gaussian. Note that x must have zero mean; if not, then its mean could be subtracted, which would reduce $\mathsf{E}\left\{\|x\|^2\right\}$ but not affect $\mathsf{I}\{y;x\}$. Consequently,

$$\mathsf{Tr}\{Q\} = \mathsf{E}\left\{\|x\|^2\right\}. \tag{C.27}$$

Maximization of the right-hand side of (C.25) with respect to Q under the constraint (C.24) gives the capacity,

$$C = \max_{\substack{Q \succeq 0 \\ \mathsf{Tr}\{Q\} \leq 1}} \log_2 \left| I_M + \rho G Q G^{\mathrm{H}} \right|. \tag{C.28}$$

Performing the optimization in (C.28) entails performing a simple procedure called *waterfilling* [3, 29]. The optimal Q is positive semi-definite, $Q \succeq 0$, and hence there exists a random vector x with covariance matrix Q. The eigenvectors of the optimal Q coincide with the right singular vectors of G or, equivalently, the eigenvectors of $G^{\mathrm{H}}G$. This means that the transmitter should transmit a superposition of signals beamformed by these singular vectors.

In order to achieve the capacity C in (C.28), the transmitter must know $G^{\mathrm{H}}G$. The optimum transmit covariance Q will change as the channel matrix changes, and the acquisition of this channel information requires some expenditure of resources. To construct a transmit scheme that works reasonably well (but is not optimal) for any G, it is common to take

$$Q = \frac{1}{K} I_K. \tag{C.29}$$

Note that Q in (C.29) satisfies the power constraint (C.24). Under the constraint that Q has the form (C.29), the capacity is

$$C = \log_2 \left| I_M + \frac{\rho}{K} G G^{\mathrm{H}} \right|. \tag{C.30}$$

To achieve a rate equal to C in (C.30), the transmitter does not need to know G, but it must still know the value of C so that it can select a codebook with an appropriate rate.

At first glance, Point-to-Point MIMO with CSI available at both ends of the link would appear extremely attractive by virtue of its optimality and simplicity, since it permits an effective diagonalization of the channel through joint application of the singular value decomposition. But on further inspection, the benefits would almost never justify the effort required to obtain the CSI. If all of the singular values of $\boldsymbol{G}$ were equal, then the optimized value of (C.28) would be equal to that of (C.30). Conversely, the regime where (C.28) most outperforms (C.30) is when $\boldsymbol{G}$ has a rank of one, in which case the MIMO link has been deployed in the most unfavorable possible propagation environment.

C.3.2 Fading MIMO Channel with Perfect CSI at the Receiver

Next we consider the case that $\boldsymbol{G}$ is random, and that the encoding at the transmitter is done over a block that spans many realizations of the channel and the noise, as in Section 2.3.3 for the scalar case. Then in (C.22), $\boldsymbol{G}$ is random and a new realization of $\boldsymbol{G}$ is drawn for each new transmitted vector $\boldsymbol{x}$. The relevant mutual information is $\mathsf{I}\{\boldsymbol{y}, \boldsymbol{G}; \boldsymbol{x}\}$ and an argument similar to that in Section C.2.5 shows that if $\boldsymbol{G}$ is known to the receiver but not to the transmitter, the capacity is

$$C = \max_{\substack{\boldsymbol{Q} \succeq \boldsymbol{0} \\ \mathsf{Tr}\{\boldsymbol{Q}\} \leq 1}} \mathsf{E}\left\{\log_2 \left|\boldsymbol{I}_M + \rho \boldsymbol{G}\boldsymbol{Q}\boldsymbol{G}^{\mathrm{H}}\right|\right\}. \tag{C.31}$$

The capacity-achieving distribution of $\boldsymbol{x}$ is circularly symmetric Gaussian with zero mean and the covariance matrix $\boldsymbol{Q}$ that solves (C.31).

Since the transmitter does not know $\boldsymbol{G}$, $\boldsymbol{Q}$ must be deterministic and not depend on $\boldsymbol{G}$. One can show that in i.i.d. Rayleigh fading, where $\{g_k^m\}$ are independent $\mathcal{CN}(0, 1)$, the optimal covariance matrix $\boldsymbol{Q}$ is given by (C.29); see [3]. Intuitively, since the singular vectors of $\boldsymbol{G}$ are isotropically distributed, the channel does not favor any direction over any other, and therefore the transmit covariance should not have any special structure either.

C.4 Multiuser MIMO Channel

The Multiuser MIMO scenario arises when a multiple-antenna base station communicates with a multiplicity of terminals simultaneously in the same time-frequency resource; see Section 1.2. Here we treat only the case of single-antenna terminals, although capacity results are known also for the general case of multiple antennas at the terminals.

There are two cases of principal interest: the uplink (see Figure 1.3(a)) and the downlink (see Figure 1.3(b)). Detailed models are given in Section 2.2.1. In information-theoretic analysis, the uplink channel is often called a *multiple-access channel* and the downlink channel is called a *broadcast channel*. We stress that in the broadcast channel, each

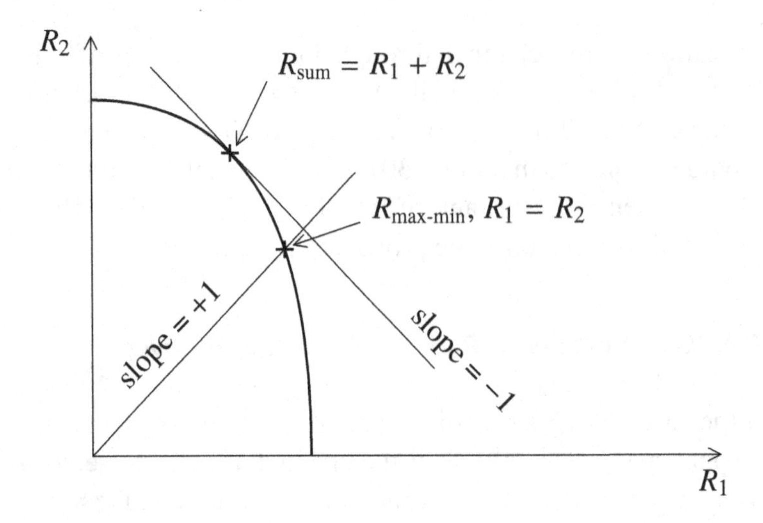

Figure C.3. Rate region for the case of two terminals, $K = 2$.

terminal receives *different* data; the special case when the same data are transmitted to all terminals is referred to as a "multicast" channel and is not of interest here. Also, in practice, there may be multiple base stations, each one serving a different set of terminals associated with it. Such a setup is known as a (MIMO) *interference channel* and is not treated here.

In both the uplink and the downlink, there are K simultaneously active communication links in each time-frequency resource. Suppose that at some operating point, the K links collectively achieve the rate K-tuple $(R_1, \ldots, R_K)$. The set of rate tuples $\{(R_1, \ldots, R_K)\}$ that is achieved for some permissible choice of operating parameters (e.g., coding scheme, and transmit powers) is called an *achievable rate region*; see Figure C.3.

In many cases, the rate region is difficult to characterize or not even of interest. One can then define a scalar measure that acts as a proxy for the whole region. Two common such measures are:

- The *sum rate*, R_{sum}. This is the largest value of $\sum_{k=1}^{K} R_k$ for any point in the region. It is clear that the corresponding rate tuple(s) $(R_1, \ldots, R_K)$ must be located on the outer boundary of the region (known as the Pareto boundary). For the special case of $K = 2$, R_{sum} corresponds to the outermost boundary point that osculates a straight line with slope -1.

- The *max-min rate*, $R_{max\text{-}min}$. This is the value of $\sum_{k=1}^{K} R_k$ at the boundary point where $R_1 = \ldots = R_K$, corresponding to a "fair" (in the egalitarian sense) operating point. In the case of $K = 2$, $R_{max\text{-}min}$ corresponds to the point where the outer boundary intersects a straight line that goes through the origin and has slope $+1$.

The largest possible rate region that can be obtained under the given power constraint (to be discussed in more detail shortly) is called the *capacity region*. The capacity region is always convex. This is so because if A and B are two points in the capacity region, then any point on the line segment $\overline{AB}$ can be expressed as $\tau A + (1-\tau)B$ for some τ, $0 \le \tau \le 1$, and can be achieved by operating at A a fraction τ of the time and at B a fraction $1-\tau$ of the time. For a given capacity region, the *sum capacity*, C_{sum}, and the *max-min capacity*, $C_{\text{max-min}}$, are defined analogously to R_{sum} and $R_{\text{max-min}}$.

An important distinction between the Point-to-Point MIMO channel (see Sections 1.1 and C.3) and the Multiuser MIMO channel is that of *cooperation*. In the point-to-point case, the antennas in the array cooperate by jointly and phase-coherently processing all signals, both at the transmitter and at the receiver. By contrast, in the multiuser case, while the base station antennas cooperate, each terminal performs coding and decoding independently; different terminals do not cooperate. Another distinction is that of power constraints: in the multiple-access (uplink) channel, each terminal is subject to an individual power constraint, whereas in the point-to-point and the broadcast (downlink) channel, the power constraint is specified in terms of the total radiated power summed over all antennas. Many equations are formally similar or identical between the Point-to-Point MIMO and Multiuser MIMO cases, but their operational meanings are different.

The following two sections are only an introduction to multiuser information theory. The intention is to give the interested reader some intuition as to how optimal performance can be achieved, and in so doing to impart first, a sense for the complexity of the subject and, second, an appreciation of the comparative simplicity of the linear precoding and decoding that Massive MIMO employs.

C.4.1 Multiple-Access Channel (Uplink)

We first consider the multiple-access channel. Referring to Section 2.2.1, the model is

$$y = \sqrt{\rho_{\text{ul}}}\boldsymbol{G}\boldsymbol{x} + \boldsymbol{w}, \tag{C.32}$$

where $\boldsymbol{x}$ is transmitted, $\boldsymbol{y}$ is received, $\boldsymbol{G}$ represents the channel, and $\boldsymbol{w}$ is noise with independent $\text{CN}(0,1)$ elements. As before, ρ_{ul} quantifies the SNR. The power constraint is defined individually for each terminal,

$$\mathsf{E}\left\{|x_k|^2\right\} \le 1, \tag{C.33}$$

for all k (inequality because the kth terminal could, in principle, opt for not spending its maximum allowed power). The constraint (C.33) can be easily modified to the case that the terminals have different power constraints – such a change could simply be absorbed by scaling the columns of $\boldsymbol{G}$.

Sum Capacity

The sum capacity is [22, 150]:

$$C_{\text{sum}} = \log_2 \left| \boldsymbol{I}_M + \rho_{\text{ul}} \boldsymbol{G}\boldsymbol{G}^{\text{H}} \right|. \tag{C.34}$$

To achieve the sum capacity in (C.34), the base station must know $\boldsymbol{G}$ and each terminal must know at what rate to transmit.

Formally, (C.34) is identical to the capacity of the Point-to-Point MIMO uplink channel, (1.1), under the constraint that all antennas emit independent streams with unit power. However, this similarity is superficial since in the model considered here, the transmit antennas (terminals) cannot cooperate, whereas in the point-to-point case they can.

Proof of (C.34) for $K = 2$

In what follows, we give a simple proof that (C.34) is the sum capacity, for the special case of $K = 2$ terminals. To simplify notation, we set $\rho_{\text{ul}} = 1$. (A different value of ρ_{ul} can be handled by simply scaling $\boldsymbol{G}$.) Suppose the terminals use Gaussian codebooks and transmit with the maximum permitted power, and that the base station operates as follows. First, x_1 is decoded by treating the quantity $\boldsymbol{w} + \boldsymbol{g}_2 x_2$ as noise, albeit correlated over its M components. We process the received signal by multiplying by a whitening operator,

$$\left(\boldsymbol{I}_M + \boldsymbol{g}_2\boldsymbol{g}_2^{\text{H}}\right)^{-1/2} \boldsymbol{y} = \left(\boldsymbol{I}_M + \boldsymbol{g}_2\boldsymbol{g}_2^{\text{H}}\right)^{-1/2} \boldsymbol{g}_1 x_1 + \boldsymbol{w}', \tag{C.35}$$

where the components of $\boldsymbol{w}'$ are independent $\mathcal{CN}(0, 1)$. Hence, we have the equivalent of a single-input multiple-output AWGN channel, a special case of the channel treated in Section C.3.1. Its rate is

$$R_1 = \log_2 \left(1 + \boldsymbol{g}_1^{\text{H}} \left(\boldsymbol{I}_M + \boldsymbol{g}_2\boldsymbol{g}_2^{\text{H}}\right)^{-1} \boldsymbol{g}_1\right). \tag{C.36}$$

The rate in (C.36) can be achieved without any knowledge of what terminal 2 transmitted; hence, whatever terminal 1 transmitted – once decoded – can be subtracted from $\boldsymbol{y}$. This results in the following interference-free effective received signal:

$$\boldsymbol{y}' = \boldsymbol{g}_2 x_2 + \boldsymbol{w}, \tag{C.37}$$

from which x_2 can be decoded by multiplying with $\boldsymbol{g}_2^{\text{H}}/\|\boldsymbol{g}_2\|^2$. The SNR in the filter output is $\|\boldsymbol{g}_2\|^2$ and hence the rate is

$$R_2 = \log_2 \left(1 + \|\boldsymbol{g}_2\|^2\right). \tag{C.38}$$

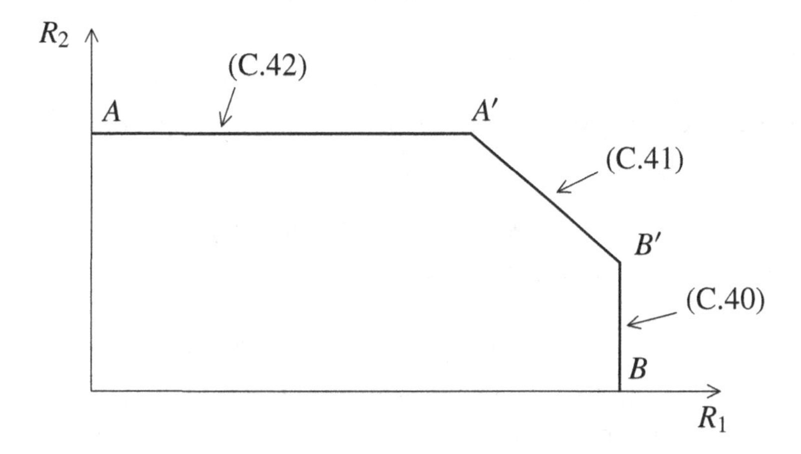

Figure C.4. Capacity region for the multiple-access channel (uplink), in the special case of $K = 2$ terminals.

We take the sum of (C.36) and (C.38) as follows:

$$
\begin{aligned}
R_1 + R_2 &= \log_2\left(1 + \|\boldsymbol{g}_2\|^2\right) + \log_2\left(1 + \boldsymbol{g}_1^{\mathrm{H}}\left(\boldsymbol{I}_M + \boldsymbol{g}_2\boldsymbol{g}_2^{\mathrm{H}}\right)^{-1}\boldsymbol{g}_1\right) \\
&\overset{(a)}{=} \log_2\left|\boldsymbol{I}_M + \boldsymbol{g}_2\boldsymbol{g}_2^{\mathrm{H}}\right| + \log_2\left|\boldsymbol{I}_M + \left(\boldsymbol{I}_M + \boldsymbol{g}_2\boldsymbol{g}_2^{\mathrm{H}}\right)^{-1/2}\boldsymbol{g}_1\boldsymbol{g}_1^{\mathrm{H}}\left(\boldsymbol{I}_M + \boldsymbol{g}_2\boldsymbol{g}_2^{\mathrm{H}}\right)^{-1/2}\right| \\
&= \log_2\left|\boldsymbol{I}_M + \boldsymbol{g}_2\boldsymbol{g}_2^{\mathrm{H}}\right| + 2\log_2\left|\left(\boldsymbol{I}_M + \boldsymbol{g}_2\boldsymbol{g}_2^{\mathrm{H}}\right)^{-1/2}\right| + \log_2\left|\boldsymbol{I}_M + \boldsymbol{g}_2\boldsymbol{g}_2^{\mathrm{H}} + \boldsymbol{g}_1\boldsymbol{g}_1^{\mathrm{H}}\right| \\
&= \log_2\left|\boldsymbol{I}_M + \boldsymbol{G}\boldsymbol{G}^{\mathrm{H}}\right|, \qquad \text{(C.39)}
\end{aligned}
$$

which is equal to C_{sum} in (C.34). In (a), we used Sylvester's determinant theorem. Hence, this strategy, known as MMSE decoding with successive interference cancellation, achieves the sum rate (C.34).

Since performance cannot be decreased by allowing the terminals to cooperate, the capacity of a Point-to-Point MIMO link with $\boldsymbol{Q} = \boldsymbol{I}_K$ provides an upper bound on the multiple-access channel sum capacity. Hence, (C.34) is the sum capacity (for $K = 2$). It should be apparent that in the above calculation the roles of the two terminals can be reversed without changing the sum rate, however, the individual rates of the terminals are changed.

Capacity Region

For $K = 2$, the capacity region has the shape of a polygon, specified by the three linear constraints,

$$R_1 \leq \log_2\left(1 + \rho_{\text{ul}}\|\boldsymbol{g}_1\|^2\right) \tag{C.40}$$

$$R_1 + R_2 \leq \log_2\left|\boldsymbol{I}_M + \rho_{\text{ul}}\boldsymbol{G}\boldsymbol{G}^{\text{H}}\right| \tag{C.41}$$

$$R_2 \leq \log_2\left(1 + \rho_{\text{ul}}\|\boldsymbol{g}_2\|^2\right), \tag{C.42}$$

as illustrated in Figure C.4. To see why the region in (C.40)–(C.42) is achievable, note first that the successive interference cancellation strategy as described above results in the point A' in Figure C.4, defined by (C.36) and (C.38). By an entirely symmetric argument, we find that the point B' is also achievable, where R_1 and R_2 are given by (C.36) and (C.38) but with $\boldsymbol{g}_1$ and $\boldsymbol{g}_2$ interchanged – here, the data of terminal 2 are the decoded first and the data of terminal 1 second. Hence, by using time-sharing, any point on the line segment $\overline{A'B'}$ can also be reached. Note that any point on this line segment achieves the sum rate C_{sum} in (C.34). Finally, it is clear that the points A and B can be achieved by activating only one of the terminals, and, hence, any points on the lines $\overline{AA'}$ and $\overline{BB'}$ are achievable too. Since any point outside of the region given by (C.40)–(C.42) would either yield a higher sum rate than C_{sum} or violate (C.40) respectively (C.42) in case only one terminal is transmitting, it is clear that (C.40)–(C.42) must be the capacity region.

C.4.2 Broadcast Channel (Downlink)

Next we consider the broadcast (downlink) channel. The model is (see Section 2.2.1),

$$\boldsymbol{y} = \sqrt{\rho_{\text{dl}}}\boldsymbol{G}^{\text{T}}\boldsymbol{x} + \boldsymbol{w}, \tag{C.43}$$

where $\boldsymbol{x}$ is transmitted, $\boldsymbol{G}$ represents the channel, $\boldsymbol{y}$ is received, ρ_{dl} is the SNR, and $\boldsymbol{w}$ is noise with independent CN(0, 1) elements. The power constraint is defined in terms of the total radiated power,

$$\mathsf{E}\left\{\|\boldsymbol{x}\|^2\right\} \leq 1. \tag{C.44}$$

Sum Capacity

The sum capacity is known to be the solution of the following optimization problem:

$$C_{\text{sum}} = \max_{\substack{\nu_k \geq 0 \\ \sum_{k=1}^{K} \nu_k \leq 1}} \log_2\left|\boldsymbol{I}_M + \rho_{\text{dl}}\boldsymbol{G}\boldsymbol{D}_\nu\boldsymbol{G}^{\text{H}}\right|, \tag{C.45}$$

where $\boldsymbol{\nu} = [\nu_1, \ldots, \nu_K]^{\mathrm{T}}$. Achieving the sum capacity in (C.45) requires that the base station and the terminals know $\boldsymbol{G}$.

Note that formally (C.45) is reminiscent of the expression (1.2) for the capacity of the Point-to-Point MIMO downlink channel, with the additional restriction that all emitted streams be independent. However, this similarity is only superficial; while (1.2) requires that the receive antennas cooperate, (C.45) does not. In fact, in the broadcast channel, the terminals cannot cooperate. Equation (C.45) is a much stronger result than (1.2).

The general, compact formula (C.45) for the sum capacity was derived in [21, 151]. Expressions for the whole capacity region are also available [152]. A result that is equivalent to (C.45) for the special case of $K = 2$ was first given in [20].

Proof of Achievability of (C.45) for $K = 2$

In what follows, we give a succinct proof of achievability in the special case of $K = 2$. We set $\rho_{\mathrm{dl}} = 1$ for simplicity; any non-unity ρ_{dl} can easily be absorbed by scaling $\boldsymbol{G}$.

Suppose the base station transmits

$$\boldsymbol{x} = \sqrt{\eta_1}\boldsymbol{a}_1 u_1 + \sqrt{\eta_2}\boldsymbol{a}_2 u_2, \tag{C.46}$$

where $\{u_1, u_2\}$ are unit-energy (independent) symbols aimed at the two terminals, $\{\boldsymbol{a}_1, \boldsymbol{a}_2\}$ are beamforming vectors (to be chosen shortly) normalized such that

$$\|\boldsymbol{a}_1\| = \|\boldsymbol{a}_2\| = 1, \tag{C.47}$$

and $\{\eta_1, \eta_2\}$ are relative powers spent on the transmission to terminal 1 and 2 respectively; $\eta_1 \geq 0$ and $\eta_2 \geq 0$. For $\boldsymbol{x}$ in (C.46) to satisfy (C.44), we require that

$$\eta_1 + \eta_2 \leq 1. \tag{C.48}$$

When encoding the data aimed at terminals 1 and 2, resulting in the symbols $\{u_1, u_2\}$, a technique known as *dirty-paper coding* [153] is used. With this technique, the information aimed at terminal 1 is coded in such a way that it can be decoded without any degradation caused by the interference from the concurrent transmission to terminal 2. The dirty-paper coding theorem that makes this possible states that the capacity of an AWGN channel, under a given power constraint, is unaffected by the presence of interference, as long as the transmitter knows this interference and can duly take it into account in the coding. If dirty-paper coding is applied, terminal 1 receives, after dirty-paper decoding,

$$y_1 = \sqrt{\eta_1}\boldsymbol{g}_1^{\mathrm{T}}\boldsymbol{a}_1 u_1 + w_1, \tag{C.49}$$

and terminal 2 receives

$$y_2 = \sqrt{\eta_1} \boldsymbol{g}_2^{\mathrm{T}} \boldsymbol{a}_1 u_1 + \sqrt{\eta_2} \boldsymbol{g}_2^{\mathrm{T}} \boldsymbol{a}_2 u_2 + w_2, \tag{C.50}$$

where w_1 and w_2 are independent CN(0, 1) noises. Note that dirty-paper coding can only be applied when encoding the information to terminal 1; operationally, in the transmitter, u_2 is decided before u_1.

The SNR in the signal seen by terminal 1 is

$$\text{SNR}_1 = \eta_1 \left| \boldsymbol{g}_1^{\mathrm{T}} \boldsymbol{a}_1 \right|^2, \tag{C.51}$$

and if interference is treated as noise, then the SINR in the signal seen by terminal 2 is

$$\text{SINR}_2 = \frac{\eta_2 \left| \boldsymbol{g}_2^{\mathrm{T}} \boldsymbol{a}_2 \right|^2}{\eta_1 \left| \boldsymbol{g}_2^{\mathrm{T}} \boldsymbol{a}_1 \right|^2 + 1}. \tag{C.52}$$

The sum rate is found by summing up the corresponding rates,

$$\begin{aligned} &\log_2 (1 + \text{SNR}_1) + \log_2 (1 + \text{SINR}_2) \\ &= \log_2 \left(1 + \eta_1 \left| \boldsymbol{g}_1^{\mathrm{T}} \boldsymbol{a}_1 \right|^2\right) + \log_2 \left(1 + \frac{\eta_2 \left| \boldsymbol{g}_2^{\mathrm{T}} \boldsymbol{a}_2 \right|^2}{\eta_1 \left| \boldsymbol{g}_2^{\mathrm{T}} \boldsymbol{a}_1 \right|^2 + 1}\right). \end{aligned} \tag{C.53}$$

We need to show that (C.53) can reach the optimum in (C.45). We accomplish this by showing that for any $\{\nu_1, \nu_2\}$ such that $\nu_1 + \nu_2 \leq 1$, there exist (i) $\{\eta_1, \eta_2\}$ such that $\eta_1 + \eta_2 \leq 1$, and (ii) vectors $\{\boldsymbol{a}_1, \boldsymbol{a}_2\}$ that satisfy (C.47) such that (C.53) is equal to the objective function in (C.45).

First, we first use a calculation entirely analogous to that in (C.39), to rewrite the objective function in (C.45) as follows:

$$\begin{aligned} &\log_2 \left| \boldsymbol{I}_M + \boldsymbol{G} \boldsymbol{D}_\nu \boldsymbol{G}^{\mathrm{H}} \right| \\ &= \log_2 \left(1 + \nu_1 \boldsymbol{g}_1^{\mathrm{H}} \left(\boldsymbol{I}_M + \nu_2 \boldsymbol{g}_2 \boldsymbol{g}_2^{\mathrm{H}}\right)^{-1} \boldsymbol{g}_1\right) + \log_2 \left(1 + \nu_2 \|\boldsymbol{g}_2\|^2\right). \end{aligned} \tag{C.54}$$

Then, we equate the SNR and SINR that appear in (C.54) and (C.53),

$$\nu_1 \boldsymbol{g}_1^{\mathrm{H}} \left(\boldsymbol{I}_M + \nu_2 \boldsymbol{g}_2 \boldsymbol{g}_2^{\mathrm{H}}\right)^{-1} \boldsymbol{g}_1 = \eta_1 \left| \boldsymbol{g}_1^{\mathrm{T}} \boldsymbol{a}_1 \right|^2, \tag{C.55}$$

$$\nu_2 \|\boldsymbol{g}_2\|^2 = \frac{\eta_2 \left| \boldsymbol{g}_2^{\mathrm{T}} \boldsymbol{a}_2 \right|^2}{\eta_1 \left| \boldsymbol{g}_2^{\mathrm{T}} \boldsymbol{a}_1 \right|^2 + 1}. \tag{C.56}$$

From (C.55) and (C.56), for any $\{\nu_1, \nu_2\}$, we have

$$\eta_1 = \frac{\nu_1 \boldsymbol{g}_1^{\mathrm{H}} \left(\boldsymbol{I}_M + \nu_2 \boldsymbol{g}_2 \boldsymbol{g}_2^{\mathrm{H}} \right)^{-1} \boldsymbol{g}_1}{\left| \boldsymbol{g}_1^{\mathrm{T}} \boldsymbol{a}_1 \right|^2}, \tag{C.57}$$

$$\eta_2 = \frac{\nu_2 \|\boldsymbol{g}_2\|^2 \left(\eta_1 \left| \boldsymbol{g}_2^{\mathrm{T}} \boldsymbol{a}_1 \right|^2 + 1 \right)}{\left| \boldsymbol{g}_2^{\mathrm{T}} \boldsymbol{a}_2 \right|^2}. \tag{C.58}$$

By selecting $\{\boldsymbol{a}_1, \boldsymbol{a}_2\}$ as follows:

$$\boldsymbol{a}_1 = \frac{\left(\boldsymbol{I}_M + \nu_2 \boldsymbol{g}_2^* \boldsymbol{g}_2^{\mathrm{T}} \right)^{-1} \boldsymbol{g}_1^*}{\left\| \left(\boldsymbol{I}_M + \nu_2 \boldsymbol{g}_2^* \boldsymbol{g}_2^{\mathrm{T}} \right)^{-1} \boldsymbol{g}_1^* \right\|}, \tag{C.59}$$

$$\boldsymbol{a}_2 = \frac{\boldsymbol{g}_2^*}{\|\boldsymbol{g}_2\|}, \tag{C.60}$$

it follows that

$$\begin{aligned}
\boldsymbol{g}_1^{\mathrm{H}} \left(\boldsymbol{I}_M + \nu_2 \boldsymbol{g}_2 \boldsymbol{g}_2^{\mathrm{H}} \right)^{-1} \boldsymbol{g}_1 &= \frac{\left| \boldsymbol{g}_1^{\mathrm{H}} \left(\boldsymbol{I}_M + \nu_2 \boldsymbol{g}_2 \boldsymbol{g}_2^{\mathrm{H}} \right)^{-1} \boldsymbol{g}_1 \right|^2}{\boldsymbol{g}_1^{\mathrm{H}} \left(\boldsymbol{I}_M + \nu_2 \boldsymbol{g}_2 \boldsymbol{g}_2^{\mathrm{H}} \right)^{-1} \boldsymbol{g}_1} \\
&= \frac{\left| \boldsymbol{g}_1^{\mathrm{H}} \left(\boldsymbol{I}_M + \nu_2 \boldsymbol{g}_2 \boldsymbol{g}_2^{\mathrm{H}} \right)^{-1} \boldsymbol{g}_1 \right|^2}{\boldsymbol{g}_1^{\mathrm{H}} \left(\boldsymbol{I}_M + \nu_2 \boldsymbol{g}_2 \boldsymbol{g}_2^{\mathrm{H}} \right)^{-1} \left(\boldsymbol{I}_M + \nu_2 \boldsymbol{g}_2 \boldsymbol{g}_2^{\mathrm{H}} \right) \left(\boldsymbol{I}_M + \nu_2 \boldsymbol{g}_2 \boldsymbol{g}_2^{\mathrm{H}} \right)^{-1} \boldsymbol{g}_1} \\
&= \frac{\left| \boldsymbol{g}_1^{\mathrm{H}} \left(\boldsymbol{I}_M + \nu_2 \boldsymbol{g}_2 \boldsymbol{g}_2^{\mathrm{H}} \right)^{-1} \boldsymbol{g}_1 \right|^2}{\left\| \left(\boldsymbol{I}_M + \nu_2 \boldsymbol{g}_2 \boldsymbol{g}_2^{\mathrm{H}} \right)^{-1} \boldsymbol{g}_1 \right\|^2 + \nu_2 \left| \boldsymbol{g}_2^{\mathrm{H}} \left(\boldsymbol{I}_M + \nu_2 \boldsymbol{g}_2 \boldsymbol{g}_2^{\mathrm{H}} \right)^{-1} \boldsymbol{g}_1 \right|^2} \\
&\overset{(a)}{=} \frac{\left| \boldsymbol{g}_1^{\mathrm{T}} \boldsymbol{a}_1 \right|^2}{1 + \nu_2 \left| \boldsymbol{g}_2^{\mathrm{T}} \boldsymbol{a}_1 \right|^2},
\end{aligned} \tag{C.61}$$

where in (a) we used (C.59). Also,

$$\|\boldsymbol{g}_2\|^2 = \left| \boldsymbol{g}_2^{\mathrm{T}} \boldsymbol{a}_2 \right|^2. \tag{C.62}$$

Upon substituting (C.61) and (C.62) into (C.57) and (C.58), we obtain

$$\eta_1 = \frac{\nu_1}{1 + \nu_2 \left| \boldsymbol{g}_2^{\mathrm{T}} \boldsymbol{a}_1 \right|^2}, \tag{C.63}$$

$$\begin{aligned} \eta_2 &= \nu_2 \left(\eta_1 \left| \boldsymbol{g}_2^{\mathrm{T}} \boldsymbol{a}_1 \right|^2 + 1 \right) \\ &\overset{(a)}{=} \frac{\nu_1 \nu_2 \left| \boldsymbol{g}_2^{\mathrm{T}} \boldsymbol{a}_1 \right|^2}{1 + \nu_2 \left| \boldsymbol{g}_2^{\mathrm{T}} \boldsymbol{a}_1 \right|^2} + \nu_2, \end{aligned} \tag{C.64}$$

where in (a) we used (C.63). Therefore, $\eta_1 + \eta_2 = \nu_1 + \nu_2$. This implies that for any $\{\nu_1, \nu_2\}$ such that $\nu_1 + \nu_2 \leq 1$, there exist $\{\eta_1, \eta_2\}$ with $\eta_1 + \eta_2 \leq 1$ such that (C.55) and (C.56) are satisfied. This concludes the proof.

Note that our demonstration that the sum rate (C.45) is achievable does *not* prove that this constitutes the maximum possible sum rate. A proof of this fact is considerably more involved.

C.4.3 Fading and Imperfect CSI

For the fading Multiuser MIMO channel, assuming that the requirements on availability of CSI as described above for a deterministic channel are satisfied, (ergodic) capacities can be defined similarly to the scalar case, by taking expectation with respect to $\boldsymbol{G}$.

If only imperfect CSI is available, then general expressions for the capacity region and the sum capacity are unknown. The linear processing bounds of Chapter 3 are exceedingly useful because they incorporate channel estimation errors, and they can closely approach the performance of the as-yet unknown optimal nonlinear processors.

Appendix D

ALTERNATIVE SINGLE-CELL CAPACITY BOUNDS

The capacity bounds derived in Chapter 3 were chosen for their utility and tractability. However, these are not the only possible bounds. In this appendix, we derive alternative bounds that, in some cases, may be tighter than their counterparts in Chapter 3.

D.1 Uplink Zero-Forcing

In the uplink, the variances of the effective noises in (3.23) and (3.32) depend on the channel estimate, which itself is known by the base station. As a result, the corresponding capacity bounds entail inconvenient expectations of logarithms of stochastic quantities. In Chapter 3, we circumvented this difficulty via the "use and forget CSI" trick, which converted the problem into one that was amenable to the bounding technique of Section 2.3.2. We point out that the same issue does not occur in the case of the downlink, where the terminal is ignorant of the channel estimate.

D.1.1 Capacity Lower Bound via Jensen's Inequality

Jensen's inequality – see Section C.1 – circumvents the difficulty of taking the expectation of a logarithm by bringing the expectation inside the logarithm. We apply Jensen in two ways: first to produce a lower bound on the expectation, second to produce an upper bound.

The application of Jensen's inequality to (3.26), in the form (C.3), yields

$$C^{\text{zf,ul}}_{\text{inst.},k} \geq \log_2\left(1 + \left(\frac{\left(1 + \rho_{\text{ul}} \sum_{k'=1}^{K} (\beta_{k'} - \gamma_{k'})\,\eta_{k'}\right) \mathsf{E}\left\{\left[\left(\mathbf{Z}^{\text{H}}\mathbf{Z}\right)^{-1}\right]_{kk}\right\}}{\rho_{\text{ul}}\gamma_k\eta_k}\right)^{-1}\right)$$

$$= \log_2\left(1 + \frac{(M-K)\rho_{\text{ul}}\gamma_k\eta_k}{1 + \rho_{\text{ul}} \sum_{k'=1}^{K} (\beta_{k'} - \gamma_{k'})\,\eta_{k'}}\right), \tag{D.1}$$

where we have used the identity (B.1). This argument merely reproduces the earlier "use and forget CSI" bound in (3.28), but it is interesting that radically different approaches yield the same bound.

D.1.2 Tightness of the Lower Bound (3.28)

We now show that the expression (3.28) is typically an excellent approximation to (3.26), by finding an *upper bound* on the expectation of the logarithm (3.26). To this end, we apply Jensen's inequality in the form (C.2) to (3.26),

$$\mathsf{E}\left\{\log_2\left(1 + \frac{\rho_{\text{ul}}\gamma_k\eta_k}{\left(1 + \rho_{\text{ul}} \sum_{k'=1}^{K} (\beta_{k'} - \gamma_{k'})\,\eta_{k'}\right)\left[\left(\mathbf{Z}^{\text{H}}\mathbf{Z}\right)^{-1}\right]_{kk}}\right)\right\}$$

$$\leq \log_2\left(1 + \frac{\rho_{\text{ul}}\gamma_k\eta_k}{1 + \rho_{\text{ul}} \sum_{k'=1}^{K} (\beta_{k'} - \gamma_{k'})\,\eta_{k'}} \mathsf{E}\left\{\frac{1}{\left[\left(\mathbf{Z}^{\text{H}}\mathbf{Z}\right)^{-1}\right]_{kk}}\right\}\right)$$

$$= \log_2\left(1 + \frac{(M+1-K)\rho_{\text{ul}}\gamma_k\eta_k}{1 + \rho_{\text{ul}} \sum_{k'=1}^{K} (\beta_{k'} - \gamma_{k'})\,\eta_{k'}}\right), \tag{D.2}$$

where we have also used the identity (B.2). Together (3.28) and (D.2) imply that

$$
\begin{aligned}
&\log_2\left(1 + \frac{(M-K)\rho_{\mathrm{ul}}\gamma_k\eta_k}{1 + \rho_{\mathrm{ul}}\sum\limits_{k'=1}^{K}(\beta_{k'} - \gamma_{k'})\,\eta_{k'}}\right) \\
&\quad \le \mathsf{E}\left\{\log_2\left(1 + \frac{\rho_{\mathrm{ul}}\gamma_k\eta_k}{\left(1 + \rho_{\mathrm{ul}}\sum\limits_{k'=1}^{K}(\beta_{k'} - \gamma_{k'})\,\eta_{k'}\right)\left[\left(\mathbf{Z}^{\mathrm{H}}\mathbf{Z}\right)^{-1}\right]_{kk}}\right)\right\} \\
&\quad \le \log_2\left(1 + \frac{(M+1-K)\rho_{\mathrm{ul}}\gamma_k\eta_k}{1 + \rho_{\mathrm{ul}}\sum\limits_{k'=1}^{K}(\beta_{k'} - \gamma_{k'})\,\eta_{k'}}\right).
\end{aligned}
\tag{D.3}
$$

If $M \gg K$, the gap between the left-hand side and the right-hand side of (D.3) is small, so both inequalities are tight. In the worst case, where $M = 2$ and $K = 1$, there is a factor-of-two gap between the lower and upper effective SINRs.

D.2 Uplink Maximum-Ratio

For uplink maximum-ratio processing, the application of Jensen's inequality produces an entirely new bound.

D.2.1 Capacity Lower Bound via Jensen's Inequality

The expression (3.35) contains two random quantities, z_k and $z_k^{\mathrm{H}} z_{k'}/\|z_k\|$ which are statistically independent. The independence follows from the fact that z_k and $z_{k'}$ are independent, the quantity $z_k/\|z_k\|$ has unit norm, and conditioned on this unit-norm vector, the quantity $z_k^{\mathrm{H}} z_{k'}/\|z_k\|$ is a $\mathrm{CN}(0, 1)$ random variable.

We apply the inequality (C.3) to (3.35),

$$
\begin{aligned}
C^{\text{mr,ul}}_{\text{inst.},k} &\geq \mathsf{E}\left\{\log_2\left(1+\frac{\rho_{\text{ul}}\gamma_k\eta_k\|z_k\|^2}{1+\rho_{\text{ul}}\sum\limits_{k'=1}^{K}(\beta_{k'}-\gamma_{k'})\eta_{k'}+\rho_{\text{ul}}\sum\limits_{\substack{k'=1\\k'\neq k}}^{K}\gamma_{k'}\eta_{k'}\left|\frac{z_k^{\text{H}}z_{k'}}{\|z_k\|}\right|^2}\right)\right\}\\
&\geq \log_2\left(1+\left(\mathsf{E}\left\{\frac{1+\rho_{\text{ul}}\sum\limits_{k'=1}^{K}(\beta_{k'}-\gamma_{k'})\eta_{k'}+\rho_{\text{ul}}\sum\limits_{\substack{k'=1\\k'\neq k}}^{K}\gamma_{k'}\eta_{k'}\left|\frac{z_k^{\text{H}}z_{k'}}{\|z_k\|}\right|^2}{\rho_{\text{ul}}\gamma_k\eta_k\|z_k\|^2}\right\}\right)^{-1}\right)\\
&\overset{(a)}{=} \log_2\left(1+\left(\frac{1+\rho_{\text{ul}}\sum\limits_{k'=1}^{K}(\beta_{k'}-\gamma_{k'})\eta_{k'}+\rho_{\text{ul}}\sum\limits_{\substack{k'=1\\k'\neq k}}^{K}\gamma_{k'}\eta_{k'}}{(M-1)\rho_{\text{ul}}\gamma_k\eta_k}\right)^{-1}\right)\\
&= \log_2\left(1+\frac{(M-1)\rho_{\text{ul}}\gamma_k\eta_k}{1+\rho_{\text{ul}}\sum\limits_{\substack{k'=1\\k'\neq k}}^{K}\beta_{k'}\eta_{k'}+\rho_{\text{ul}}(\beta_k-\gamma_k)\eta_k}\right),
\end{aligned}
\tag{D.4}
$$

where in (a) we exploited the independence between z_k and $z_k^{\text{H}}z_{k'}/\|z_k\|$, and the fact that $\mathsf{E}\left\{\|z_k\|^{-2}\right\} = 1/(M-1)$; see (B.1).

D.2.2 Comparison of the Bounds (3.41) and (D.4)

An inspection of the two bounds (3.41) and (D.4) discloses that the latter bound has smaller coherent beamforming gain than the former, but reduced intra-cell interference. In the following, we assume that the same power control coefficients are used within the two bounds. We let

$$\mathrm{SINR}_{(3.41)} = \frac{M\rho_{\mathrm{ul}}\gamma_k\eta_k}{1+\rho_{\mathrm{ul}}\sum_{k'=1}^{K}\beta_{k'}\eta_{k'}}, \quad \text{and}$$

$$\mathrm{SINR}_{(\mathrm{D}.4)} = \frac{(M-1)\rho_{\mathrm{ul}}\gamma_k\eta_k}{1+\rho_{\mathrm{ul}}\sum_{\substack{k'=1\\k'\neq k}}^{K}\beta_{k'}\eta_{k'}+\rho_{\mathrm{ul}}\left(\beta_k-\gamma_k\right)\eta_k} \tag{D.5}$$

be the effective SINRs in (3.41) respectively (D.4). Then we have that

$$\mathrm{SINR}_{(\mathrm{D}.4)} - \mathrm{SINR}_{(3.41)} = \frac{\rho_{\mathrm{ul}}\gamma_k\eta_k}{1+\rho_{\mathrm{ul}}\sum_{k'=1}^{K}\beta_{k'}\eta_{k'}-\rho_{\mathrm{ul}}\gamma_k\eta_k}\left(\mathrm{SINR}_{(3.41)}-1\right). \tag{D.6}$$

Hence, (D.4) is a better capacity lower bound than (3.41) precisely when $\mathrm{SINR}_{(3.41)} > 1$.

To determine the possible gap between the two bounds, we divide both sides of (D.6) by $\mathrm{SINR}_{(3.41)}$,

$$\begin{aligned}\frac{\mathrm{SINR}_{(\mathrm{D}.4)}}{\mathrm{SINR}_{(3.41)}} &= 1+\frac{\rho_{\mathrm{ul}}\gamma_k\eta_k}{1+\rho_{\mathrm{ul}}\sum_{k'=1}^{K}\beta_{k'}\eta_{k'}-\rho_{\mathrm{ul}}\gamma_k\eta_k}\left(1-\frac{1}{\mathrm{SINR}_{(3.41)}}\right)\\ &< 1+\frac{\rho_{\mathrm{ul}}\gamma_k\eta_k}{1+\rho_{\mathrm{ul}}\sum_{k'=1}^{K}\beta_{k'}\eta_{k'}-\rho_{\mathrm{ul}}\gamma_k\eta_k}.\end{aligned} \tag{D.7}$$

If $\mathrm{SINR}_{(\mathrm{D}.4)} > \mathrm{SINR}_{(3.41)}$, we have

$$\begin{aligned}0 &< \log_2(1+\mathrm{SINR}_{(\mathrm{D}.4)}) - \log_2(1+\mathrm{SINR}_{(3.41)})\\ &= \log_2\left(\frac{1+\mathrm{SINR}_{(\mathrm{D}.4)}}{1+\mathrm{SINR}_{(3.41)}}\right)\\ &< \log_2\left(\frac{\mathrm{SINR}_{(\mathrm{D}.4)}}{\mathrm{SINR}_{(3.41)}}\right)\\ &< \log_2\left(1+\frac{\rho_{\mathrm{ul}}\gamma_k\eta_k}{1+\rho_{\mathrm{ul}}\sum_{k'=1}^{K}\beta_{k'}\eta_{k'}-\rho_{\mathrm{ul}}\gamma_k\eta_k}\right).\end{aligned} \tag{D.8}$$

Conversely, if $\mathrm{SINR}_{(D.4)} \leq \mathrm{SINR}_{(3.41)}$, from (D.6) we know that $\mathrm{SINR}_{(3.41)} \leq 1$. Thus,

$$\begin{aligned} 0 &< \log_2(1 + \mathrm{SINR}_{(3.41)}) - \log_2(1 + \mathrm{SINR}_{(D.4)}) \\ &< \log_2(1 + \mathrm{SINR}_{(3.41)}) \\ &\leq \log_2(2) \\ &= 1. \end{aligned} \tag{D.9}$$

The combination of (D.8) and (D.9) yields

$$\begin{aligned} &|\log_2(1 + \mathrm{SINR}_{(3.41)}) - \log_2(1 + \mathrm{SINR}_{(D.4)})| \\ &\qquad < \max\left\{1, \log_2\left(1 + \frac{\rho_{\mathrm{ul}}\gamma_k\eta_k}{1 + \rho_{\mathrm{ul}}\sum\limits_{k'=1}^{K}\beta_{k'}\eta_{k'} - \rho_{\mathrm{ul}}\gamma_k\eta_k}\right)\right\}. \end{aligned} \tag{D.10}$$

This shows that for any $M > 1$, the difference between the capacity bounds (3.41) and (D.4) is less than a constant that does not depend on M.

Interestingly, even though the difference between the two capacity bounds is bounded, the difference between the corresponding SINRs can be arbitrarily large. In fact, from (D.6), we see that the difference $\mathrm{SINR}_{(D.4)} - \mathrm{SINR}_{(3.41)}$ can become arbitrarily large as M grows, since $\mathrm{SINR}_{(3.41)} \to \infty$ when $M \to \infty$.

D.3 Downlink Maximum-Ratio

For maximum-ratio processing in the downlink, it is possible to use an alternative normalization when constructing the precoding matrix $\boldsymbol{A}$. In particular, let

$$\boldsymbol{a}_k = \sqrt{M-1}\frac{\boldsymbol{z}_k^*}{\|\boldsymbol{z}_k\|^2} \tag{D.11}$$

in lieu of (3.57). Using (B.1) in Appendix B, we find that $\mathsf{E}\left\{\|\boldsymbol{a}_k\|^2\right\} = 1$, so the power constraint $\mathsf{E}\left\{\|\boldsymbol{x}\|^2\right\} = \sum_{k=1}^K \eta_k$ is satisfied. A now-familiar derivation from Chapter 3 yields a new effective SINR within the capacity bound (3.63) but with $\mathrm{SINR}_k^{\mathrm{mr,dl}}$ replaced with

$$\left.\mathrm{SINR}_k^{\mathrm{mr,dl}}\right|_{\mathrm{alternative}} = \frac{(M-1)\rho_{\mathrm{dl}}\gamma_k\eta_k}{1 + \rho_{\mathrm{dl}}\beta_k\sum\limits_{\substack{k'=1\\k'\neq k}}^{K}\eta_{k'} + \rho_{\mathrm{dl}}(\beta_k - \gamma_k)\eta_k}. \tag{D.12}$$

Upon comparing (D.12) to (3.64), we see that this new bound reduces the coherent beamforming gain proportionately from M to $M - 1$, and reduces the beamforming gain uncertainty from $\rho_{\text{dl}}\beta_k\eta_k$ to $\rho_{\text{dl}}(\beta_k - \gamma_k)\eta_k$. In practice, the difference in performance between these two variants of maximum-ratio processing is marginal.

The new equivalent SINR (D.12) eliminates a discrepancy between the $K = 1$ zero-forcing SINR and the $K = 1$ maximum-ratio SINR seen in Table 3.1, since, for the special case of $K = 1$, the normalization (D.11) is mathematically equivalent to zero-forcing.

There are several additional, nearly equivalent, possibilities to normalize the precoder $\boldsymbol{A}$ – for example, dividing by a constant proportional to $||z_k||$, $\text{Tr}\left\{\boldsymbol{Z}\boldsymbol{Z}^{\text{H}}\right\}$ or $\sqrt{\text{Tr}\left\{\boldsymbol{Z}\boldsymbol{Z}^{\text{H}}\right\}}$ instead of $||z_k||^2$.

The above bounding technique can be applied to the uplink by employing the modified decoder, $\boldsymbol{a}_k = z_k/||z_k||^2$, combined with the "use and forget CSI" trick. The resulting bound, however, turns out to be identical to (D.4).

Appendix E

ASYMPTOTIC SINR IN MULTI-CELL SYSTEMS

Reference [26] took a non-Bayesian approach to precoding and decoding that requires no knowledge of the large-scale fading. On the uplink, every terminal transmits data at full power, so $\eta_{l'k} = 1$, which yields the following asymptotic SINR:

$$
\begin{aligned}
\lim_{M\to\infty} \mathrm{SINR}^{\mathrm{ul}}_{lk} &= \frac{\gamma^{l}_{lk}}{\sum\limits_{l'\in\mathcal{P}_l\setminus\{l\}} \gamma^{l}_{l'k}} \\
&\overset{(a)}{=} \frac{\dfrac{\tau_{\mathrm{p}}\rho_{\mathrm{ul}}\left(\beta^{l}_{lk}\right)^2}{1+\tau_{\mathrm{p}}\rho_{\mathrm{ul}}\sum\limits_{l''\in\mathcal{P}_l}\beta^{l}_{l''k}}}{\sum\limits_{l'\in\mathcal{P}_l\setminus\{l\}} \dfrac{\tau_{\mathrm{p}}\rho_{\mathrm{ul}}\left(\beta^{l}_{l'k}\right)^2}{1+\tau_{\mathrm{p}}\rho_{\mathrm{ul}}\sum\limits_{l''\in\mathcal{P}_l}\beta^{l}_{l''k}}} \\
&= \frac{\left(\beta^{l}_{lk}\right)^2}{\sum\limits_{l'\in\mathcal{P}_l\setminus\{l\}}\left(\beta^{l}_{l'k}\right)^2},
\end{aligned}
\tag{E.1}
$$

where, in (a), we used (4.4).

On the downlink, the de-spread pilot signals, (4.1) are used directly for maximum-ratio processing, so implicitly the power control coefficient is proportional to the variance of the de-spread pilot signal,

$$\begin{aligned}\eta_{l'k} &\propto \mathsf{E}\left\{\left\|\left[\boldsymbol{Y}'_{\mathrm{p}l'}\right]_k\right\|^2\right\} \\ &= 1 + \tau_{\mathrm{p}}\rho_{\mathrm{ul}} \sum_{l''\in\mathcal{P}'_l} \beta^{l'}_{l''k}.\end{aligned} \tag{E.2}$$

This gives the following asymptotic SINR:

$$\begin{aligned}\lim_{M\to\infty} \mathrm{SINR}^{\mathrm{dl}}_{lk} &= \frac{\gamma^l_{lk}\eta_{lk}}{\sum\limits_{l'\in\mathcal{P}_l\setminus\{l\}} \gamma^{l'}_{lk}\eta_{l'k}} \\ &= \frac{\gamma^l_{lk}\left(1 + \tau_{\mathrm{p}}\rho_{\mathrm{ul}} \sum\limits_{l''\in\mathcal{P}_l} \beta^l_{l''k}\right)}{\sum\limits_{l'\in\mathcal{P}_l\setminus\{l\}} \gamma^{l'}_{lk}\left(1 + \tau_{\mathrm{p}}\rho_{\mathrm{ul}} \sum\limits_{l''\in\mathcal{P}'_l} \beta^{l'}_{l''k}\right)} \\ &= \frac{\left[\dfrac{\tau_{\mathrm{p}}\rho_{\mathrm{ul}}\left(\beta^l_{lk}\right)^2}{1 + \tau_{\mathrm{p}}\rho_{\mathrm{ul}} \sum\limits_{l''\in\mathcal{P}_l} \beta^l_{l''k}}\right]\left(1 + \tau_{\mathrm{p}}\rho_{\mathrm{ul}} \sum\limits_{l''\in\mathcal{P}_l} \beta^l_{l''k}\right)}{\sum\limits_{l'\in\mathcal{P}_l\setminus\{l\}}\left[\dfrac{\tau_{\mathrm{p}}\rho_{\mathrm{ul}}\left(\beta^{l'}_{lk}\right)^2}{1 + \tau_{\mathrm{p}}\rho_{\mathrm{ul}} \sum\limits_{l''\in\mathcal{P}_{l'}} \beta^{l'}_{l''k}}\right]\left(1 + \tau_{\mathrm{p}}\rho_{\mathrm{ul}} \sum\limits_{l''\in\mathcal{P}'_l} \beta^{l'}_{l''k}\right)} \\ &= \frac{\left(\beta^l_{lk}\right)^2}{\sum\limits_{l'\in\mathcal{P}_l\setminus\{l\}} \left(\beta^{l'}_{lk}\right)^2}.\end{aligned} \tag{E.3}$$

The SINRs in (E.1) and (E.3) are equal to those obtained in [26].

Appendix F

LINK BUDGET CALCULATIONS

The nominal uplink SNR ρ_{ul}, in dB, for a terminal with unit pathloss ($\beta = 1$) can be calculated as follows:

$$\begin{aligned}\rho_{\text{ul}} = & \text{ [uplink radiated power (dBm)]} \\ & + \text{[antenna gain (dB)]}_{\text{base station}} \\ & + \text{[antenna gain (dB)]}_{\text{terminal}} \\ & - \text{[effective noise power at the base station receiver (dBm)]},\end{aligned} \tag{F.1}$$

where

$$\begin{aligned}\text{[effective noise power at the base station (dBm)]} = & \ [BN_0 \text{ (dBm)}]_{\text{base station}} \\ & + \text{[noise figure (dB)]}_{\text{base station}}.\end{aligned} \tag{F.2}$$

In (F.2), BN_0 is the receiver noise power, where

$$N_0 = k_B T_{\text{noise}} \tag{F.3}$$

is the noise power spectral density, B is the system bandwidth in Hz,

$$k_B \approx 1.38 \times 10^{-23} \qquad \text{(Joule/Kelvin)} \tag{F.4}$$

is the Boltzmann constant and T_{noise} is the noise temperature in Kelvin.

As an example, with a system bandwidth of $B = 20$ MHz, an uplink radiated power of 1 W, a base station receiver noise figure of 9 dB, a noise temperature of 300 K, a 6 dBi terminal antenna gain, and a 0 dBi base station antenna gain, we have $\rho_{\text{ul}} = 128$ dB.

The nominal downlink SNR ρ_{dl} is calculated analogously.

Appendix G

UNIFORMLY DISTRIBUTED POINTS IN A HEXAGON

This appendix shows how to generate uniformly distributed points in a hexagon. This method is used in the numerical results in Chapter 6.

As shown in Figure G.1, a hexagon consists of three rhombuses – labeled 1, 2, and 3 in the figure. These rhombuses can be obtained by mapping the unit square in the (u, v)-plane onto the (x, y)-plane using the following three linear mappings:

$$\mathcal{F}_1 : \begin{cases} x = \frac{\sqrt{3}}{2}u \\ y = -\frac{1}{2}u + v \end{cases}, \quad \mathcal{F}_2 : \begin{cases} x = -\frac{\sqrt{3}}{2}u + \frac{\sqrt{3}}{2}v \\ y = -\frac{1}{2}u - \frac{1}{2}v \end{cases}, \text{ and } \quad \mathcal{F}_3 : \begin{cases} x = -\frac{\sqrt{3}}{2}v \\ y = u - \frac{1}{2}v \end{cases}. \quad \text{(G.1)}$$

Since a linear mapping has a constant Jacobian, a point (u, v) that is uniformly distributed in the unit square will be mapped onto a point (x, y) that is uniformly distributed in each rhombus. Hence, K uniformly distributed points in a hexagon can be generated as follows:

1. Generate K uniformly distributed points in the interval $(0, 1)$.
2. Let K_1 be the number of points in the interval $(0, 1/3)$, let K_2 be the number of points in the interval $[1/3, 2/3)$, and let K_3 be the number of points in the interval $[2/3, 1)$. We have $K_1 + K_2 + K_3 = K$.
3. For $n = 1, 2, 3$, generate K_n uniformly distributed points in the unit square, and map them to rhombus n via the linear mapping $\mathcal{F}_n$.

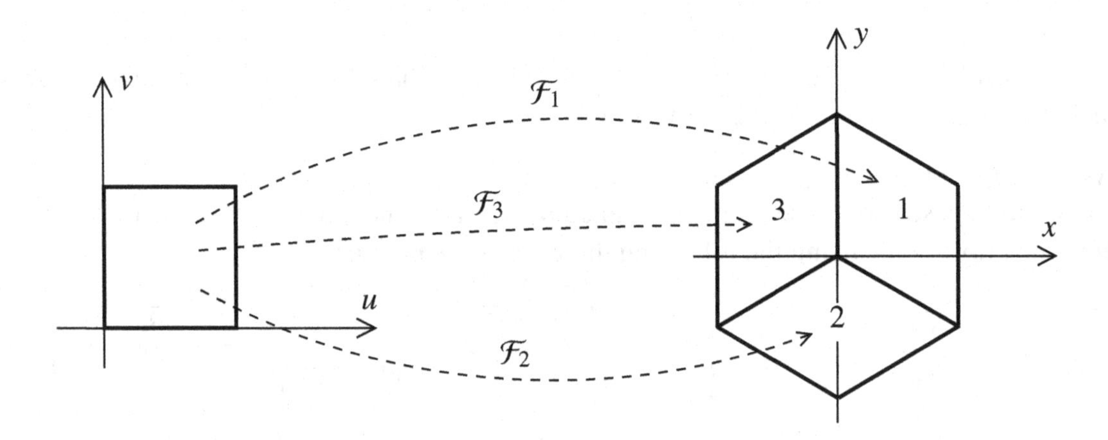

Figure G.1. Linear mappings from the unit square to the rhombuses that constitute a hexagon.

Appendix H

SUMMARY OF ABBREVIATIONS AND NOTATION

Abbreviations

AWGN	additive white Gaussian noise
CSI	channel state information
DL (dl)	downlink
FDD	frequency-division duplexing
i.i.d.	independent and identically distributed
LoS	line-of-sight
MIMO	multiple-input multiple-output
MMSE	minimum mean-square error
MR (mr)	maximum-ratio
OFDM	orthogonal frequency-division multiplexing
QAM	quadrature amplitude modulation
QPSK	quadrature phase shift keying
SINR	signal-to-interference-plus-noise ratio
SNR	signal-to-noise ratio
TDD	time-division duplexing
UL (ul)	uplink
UR-LoS	uniformly random line-of-sight
ZF (zf)	zero-forcing

Mathematical Notation and Operators

i	$\sqrt{-1}$
$(\cdot)^*$	complex conjugate
$(\cdot)^{\mathrm{T}}$	transpose
$(\cdot)^{\mathrm{H}}$	conjugate transpose
$[\cdot]_k$	the kth element of a vector
$[\cdot]_{kl}$	the (k,l)th element of a matrix
$\boldsymbol{I}_N$	$N \times N$ identity matrix
$\boldsymbol{D}_{\boldsymbol{x}}$	diagonal matrix with the elements of $\boldsymbol{x}$ on its diagonal
$p_x(\cdot)$	probability density function of random variable x
$p_{x\mid y}(\cdot\mid\cdot)$	probability density function of random variable x, conditioned on y
$\mathsf{E}\{x\}$	expected value of the random variable x
$\mathsf{E}\{x\mid y\}$	expected value of the random variable x, conditioned on y
$\mathsf{Var}\{x\}$	variance of the random variable x, $\mathsf{E}\left\{\lvert x - \mathsf{E}\{x\}\rvert^2\right\}$
$\mathsf{Var}\{x\mid y\}$	variance of x conditioned on y, $\mathsf{E}\left\{\lvert x - \mathsf{E}\{x\mid y\}\rvert^2 \middle\vert y\right\}$
$\mathsf{Cov}\{\boldsymbol{x}\}$	covariance matrix of the random vector $\boldsymbol{x}$
$\mathsf{Cov}\{\boldsymbol{x}\mid\boldsymbol{y}\}$	covariance matrix of $\boldsymbol{x}$ conditioned on $\boldsymbol{y}$
$\lvert x\rvert$	absolute value of the scalar x
$\lVert\boldsymbol{x}\rVert$	Euclidean norm of the vector $\boldsymbol{x}$
$\lvert\boldsymbol{X}\rvert$	determinant of the square matrix $\boldsymbol{X}$
$\mathsf{Tr}\{\boldsymbol{X}\}$	trace of the square matrix $\boldsymbol{X}$
$\mathrm{sinc}(x)$	(normalized) sinc function: $\mathrm{sinc}(x) = \sin(\pi x)/(\pi x)$
$\mathsf{h}\{x\}$	differential entropy of x
$\mathsf{h}\{y\mid x\}$	conditional differential entropy of y, given x
$\mathsf{I}\{y;x\}$	mutual information between x and y
χ^2_N	chi-square random variable with N degrees of freedom
$\mathrm{CN}(0,\sigma^2)$	circularly sym. Gaussian variable with zero mean and variance σ^2
$\mathrm{CN}(\boldsymbol{0},\boldsymbol{\Lambda})$	circularly sym. Gaussian vector with zero mean and covariance $\boldsymbol{\Lambda}$
$\binom{N}{n}$	binomial coefficient: $\binom{N}{n} = \frac{N!}{(N-n)!n!}$
$\boldsymbol{X} \geq \boldsymbol{0}$	the matrix $\boldsymbol{X}$ is positive semi-definite

Symbols With a Specific Meaning Throughout the Book

M	number of base station antennas
m	antenna index
K	number of terminals[1]
k, k', k''	terminal index
T_{c}	coherence time (seconds)
B_{c}	coherence bandwidth (Hz)
τ_{c}	length of coherence interval in samples
τ_{p}	number of pilot samples per coherence interval
τ_{ul}	number of uplink payload samples per coherence interval
τ_{dl}	number of downlink payload samples per coherence interval
ρ_{ul}	uplink SNR
ρ_{dl}	downlink SNR
f_{c}	carrier frequency
λ	carrier wavelength
c	speed of light (3×10^8 m/s)

Single-Cell Systems

h_k^m	small-scale fading between terminal k and base station antenna m
β_k	large-scale fading between terminal k and the base station
g_k^m	channel between terminal k and base station antenna m
$\boldsymbol{g}_k$	channel vector between the kth terminal and the base station
$\boldsymbol{G}$	channel matrix between all terminals and the base station
$\hat{g}_k^m$	estimate of the channel g_k^m
$\hat{\boldsymbol{g}}_k$	estimate of the channel $\boldsymbol{g}_k$
$\hat{\boldsymbol{G}}$	estimate of the channel $\boldsymbol{G}$
$\boldsymbol{Z}$	normalized channel estimate $\hat{\boldsymbol{G}}$
$\tilde{g}_k^m$	channel estimation error, defined as $\hat{g}_k^m - g_k^m$
$\tilde{\boldsymbol{g}}_k$	channel estimation error, defined as $\hat{\boldsymbol{g}}_k - \boldsymbol{g}_k$
$\tilde{\boldsymbol{G}}$	channel estimation error, defined as $\hat{\boldsymbol{G}} - \boldsymbol{G}$
γ_k	mean-square channel estimate $\hat{g}_m^k$ (independent of m)
η_k	power control coefficient for the kth terminal
SINR_k	effective SINR for the kth terminal

[1]Exceptionally, in Sections 1.1 and C.3, K represents the number of antennas at the terminal.

$\overline{\text{SINR}}$	common SINR with max-min fairness power control

Multi-Cell Systems

L	number of cells
l, l', l''	cell index
n_{reuse}	pilot reuse factor
$\mathcal{P}_l$	indices of the cells that use same pilot sequences as the lth cell (incl. cell l)
$h_{l'k}^{lm}$	small-scale fading between terminal k in cell l' and antenna m in cell l
$\beta_{l'k}^{l}$	large-scale fading between terminal k in cell l' and the base station in cell l
$g_{l'k}^{lm}$	channel between terminal k in cell l' and antenna m in cell l
$\boldsymbol{g}_{l'k}^{l}$	channel between terminal k in cell l' and base station l
$\boldsymbol{G}_{l'}^{l}$	channel between the terminals in cell l' and base station l
$\hat{g}_{l'k}^{lm}$	estimate of the channel $g_{l'k}^{lm}$
$\hat{\boldsymbol{g}}_{l'k}^{l}$	estimate of the channel $\boldsymbol{g}_{l'k}^{l}$
$\hat{\boldsymbol{G}}_{l'}^{l}$	estimate of the channel $\boldsymbol{G}_{l'}^{l}$
$\tilde{g}_{l'k}^{lm}$	channel estimation error, defined as $\hat{g}_{l'k}^{lm} - g_{l'k}^{lm}$
$\tilde{\boldsymbol{g}}_{l'k}^{l}$	channel estimation error, defined as $\hat{\boldsymbol{g}}_{l'k}^{l} - \boldsymbol{g}_{l'k}^{l}$
$\tilde{\boldsymbol{G}}_{l'}^{l}$	channel estimation error, defined as $\hat{\boldsymbol{G}}_{l'}^{l} - \boldsymbol{G}_{l'}^{l}$
$\boldsymbol{Z}^{l}$	normalized channel estimate $\hat{\boldsymbol{G}}_{l'}^{l}$ (independent of l')
$\gamma_{l'k}^{l}$	mean-square channel estimate $g_{l'k}^{lm}$ (independent of m)
η_{lk}	power-control coefficient for the kth terminal in the lth cell
SINR_{lk}	effective SINR for the kth terminal in the lth cell
$\overline{\text{SINR}}_{l}$	common SINR with max-min fairness power control, in the lth cell

REFERENCES

[1] E. F. W. Alexanderson, "Transoceanic radio communication," Presented at a joint meeting of the American Institute of Electrical Engineers and the Institute of Radio Engineers, New York, NY, Oct. 1919.

[2] G. J. Foschini and M. J. Gans, "On limits of wireless communications in a fading environment when using multiple antennas," *Wireless Pers. Commun.*, vol. 6, pp. 311–335, Mar. 1998.

[3] E. Telatar, "Capacity of multi-antenna Gaussian channels," *European Trans. on Telecommun.*, vol. 10, no. 6, pp. 585–596, Nov. 1999.

[4] G. G. Raleigh and J. M. Cioffi, "Spatio-temporal coding for wireless communication," *IEEE Trans. Commun.*, vol. 46, no. 3, pp. 357–366, Mar. 1998.

[5] E. G. Larsson and P. Stoica, *Space-Time Block Coding for Wireless Communications.* Cambridge University Press, 2003.

[6] J. C. Guey, M. P. Fitz, M. R. Bell, and W. Y. Kuo, "Signal design for transmitter diversity wireless communication systems over Rayleigh fading channels," *IEEE Trans. Commun.*, vol. 47, no. 4, pp. 527–537, Apr. 1999.

[7] V. Tarokh, H. Jafarkhani, and A. R. Calderbank, "Space-time block codes from orthogonal designs," *IEEE Trans. Inf. Theory*, vol. 45, no. 5, pp. 1456–1467, July 1999.

[8] V. Tarokh, N. Seshadri, and A. R. Calderbank, "Space-time codes for high data rate wireless communications: performance criterion and code construction," *IEEE Trans. Inf. Theory*, vol. 44, no. 2, pp. 744–765, Mar. 1998.

[9] S. M. Alamouti, "A simple transmit diversity technique for wireless communications," *IEEE J. Sel. Areas Commun.*, vol. 16, no. 8, pp. 1451–1458, Oct. 1998.

[10] B. M. Hochwald and S. ten Brink, "Achieving near-capacity on a multiple-antenna channel," *IEEE Trans. Commun.*, vol. 51, no. 3, pp. 389–399, Mar. 2003.

[11] T. L. Marzetta and B. M. Hochwald, "Capacity of a mobile multiple-antenna communication link in Rayleigh flat fading," *IEEE Trans. Inf. Theory*, vol. 45, no. 1, pp. 139–157, Jan. 1999.

[12] J. H. Winters, "Optimum combining in digital mobile radio with cochannel interference," *IEEE J. Sel. Areas Commun.*, vol. 2, no. 4, pp. 528–539, July 1984.

[13] ——, "Optimum combining for indoor radio systems with multiple users," *IEEE Trans. Commun.*, vol. 35, no. 11, pp. 1222–1230, Nov. 1987.

[14] S. C. Swales, M. A. Beach, D. J. Edwards, and J. P. McGeehan, "The performance enhancement of multibeam adaptive base-station antennas for cellular land mobile radio systems," *IEEE Trans. Veh. Technol.*, vol. 39, no. 1, pp. 56–67, Feb. 1990.

[15] R. Roy and B. Ottersten, "Spatial division multiple access wireless communication systems," US Patent No. 5,515,378, 1991.

[16] B. V. Veen and K. Buckley, "Beamforming: A versatile approach to spatial filtering," *IEEE ASSP Mag.*, vol. 5, no. 2, pp. 4–24, Apr. 1988.

[17] S. Anderson, M. Millnert, M. Viberg, and B. Wahlberg, "An adaptive array for mobile communication systems," *IEEE Trans. Veh. Technol.*, vol. 40, no. 1, pp. 230–236, Feb. 1991.

[18] A. Paulraj and T. Kailath, "Increasing capacity in wireless broadcast systems using distributed transmission/directional reception (DTDR)," US Patent No. 5,345,599, 1994.

[19] S. Anderson, U. Forssén, J. Karlsson, T. Witzschel, P. Fischer, and A. Krug, "Ericsson/Mannesmann GSM field-trials with adaptive antennas," in *Proc. IEE Colloquium on Advanced TDMA Techniques and Applications*, London, UK, Oct. 1996.

[20] G. Caire and S. Shamai, "On the achievable throughput of a multi-antenna Gaussian broadcast channel," *IEEE Trans. Inf. Theory*, vol. 49, no. 7, pp. 1691–1706, July 2003.

[21] P. Viswanath and D. Tse, "Sum capacity of the vector Gaussian broadcast channel and uplink-downlink duality," *IEEE Trans. Inf. Theory*, vol. 49, no. 8, pp. 1912–1921, Aug. 2003.

[22] A. Goldsmith, S. A. Jafar, N. Jindal, and S. Vishwanath, "Capacity limits of MIMO channels," *IEEE J. Sel. Areas Commun.*, vol. 21, no. 5, pp. 684–702, June 2003.

[23] S. Vishwanath, N. Jindal, and A. Goldsmith, "Duality, achievable rates, and sum-rate capacity of Gaussian MIMO broadcast channels," *IEEE Trans. Inf. Theory*, vol. 49, no. 10, pp. 2658–2668, Oct. 2003.

[24] D. Gesbert, M. Kountouris, R. W. Heath, Jr., C.-B. Chae, and T. Sälzer, "From single-user to multi-user communications: shifting the MIMO paradigm," *IEEE Signal Process. Mag.*, vol. 24, no. 5, pp. 36–46, Sept. 2007.

[25] T. L. Marzetta, "How much training is required for multiuser MIMO?" in *Proc. 40th Asilomar Conference on Signals, Systems and Computers (ACSSC)*, Pacific Grove, CA, USA, Nov. 2006.

[26] ——, "Noncooperative cellular wireless with unlimited numbers of base station antennas," *IEEE Trans. Wireless Commun.*, vol. 9, no. 11, pp. 3590–3600, Nov. 2010.

[27] T. J. Richardson, M. A. Shokrollahi, and R. L. Urbanke, "Design of capacity-approaching irregular low-density parity-check codes," *IEEE Trans. Inf. Theory*, vol. 47, no. 2, pp. 619–637, 2001.

[28] T. L. Marzetta and B. M. Hochwald, "Fast transfer of channel state information in wireless systems," *IEEE Trans. Signal Process.*, vol. 54, no. 4, pp. 1268–1278, Apr. 2006.

[29] D. Tse and P. Viswanath, *Fundamentals of Wireless Communication*. Cambridge University Press, 2005.

[30] E. Björnson, M. Bengtsson, and B. Ottersten, "Optimal multiuser transmit beamforming: A difficult problem with a simple solution structure," *IEEE Signal Process. Mag.*, vol. 31, no. 4, pp. 142–148, July 2014.

[31] T. S. Rappaport, *Wireless Communications*. Prentice-Hall, 2002.

[32] V. H. M. Donald, "The cellular concept," *Bell Labs Technical Journal*, vol. 58, no. 1, pp. 15–41, Jan. 1979.

[33] D. C. Cox, "Cochannel interference considerations in frequency reuse small-coverage-area radio systems," *IEEE Trans. Commun.*, vol. 30, no. 1, pp. 135–142, Jan. 1982.

[34] A. Gamst, "Homogeneous distribution of frequencies in a regular hexagonal cell system," *IEEE Trans. Veh. Technol.*, vol. 31, no. 3, pp. 132–144, Aug. 1982.

[35] E. Björnson, E. G. Larsson, and T. L. Marzetta, "Massive MIMO: 10 myths and one critical question," *IEEE Commun. Mag.*, vol. 54, no. 2, pp. 114–123, Feb. 2016.

[36] 3GPP, *Digital cellular telecommunications system (Phase 2+); Radio network planning aspects.* 3GPP ETSI TR, 2010.

[37] A. L. Moustakas, H. U. Baranger, L. Balents, A. M. Sengupta, and S. H. Simon, "Communication through a diffusive medium: Coherence and capacity," *Science*, vol. 287, no. 5451, pp. 287–290, 2000.

[38] R. Couillet and M. Debbah, *Random Matrix Methods for Wireless Communications.* Cambridge University Press, 2011.

[39] H. Q. Ngo, E. G. Larsson, and T. L. Marzetta, "Aspects of favorable propagation in massive MIMO," in *Proc. European Signal Processing Conf. (EUSIPCO)*, Lisbon, Portugal, Sept. 2014.

[40] A. M. Sayeed, "Deconstructing multi-antenna fading channels," *IEEE Trans. Signal Process.*, pp. 2563–2579, 2002.

[41] W. Feller, *An Introduction to Probability Theory and Its Applications*, 2nd edn. New York: Wiley, 1957, vol. 1.

[42] X. Gao, O. Edfors, F. Rusek, and F. Tufvesson, "Massive MIMO performance evaluation based on measured propagation data," *IEEE Trans. Wireless Commun.*, vol. 14, no. 7, pp. 3899–3911, July 2015.

[43] J. Hoydis, C. Hoek, T. Wild, and S. ten Brink, "Channel measurements for large antenna arrays," in *Proc. Int. Symp. of Wireless Communication Systems (ISWCS)*, Paris, France, Aug. 2012.

[44] F. Rusek, D. Persson, B. K. Lau, E. G. Larsson, T. L. Marzetta, O. Edfors, and F. Tufvesson, "Scaling up MIMO: Opportunities and challenges with very large arrays," *IEEE Signal Process. Mag.*, vol. 30, no. 1, pp. 40–60, Jan. 2013.

[45] E. G. Larsson, F. Tufvesson, O. Edfors, and T. L. Marzetta, "Massive MIMO for next generation wireless systems," *IEEE Commun. Mag.*, vol. 52, no. 2, pp. 186–195, Feb. 2014.

[46] A. O. Martinez, E. D. Carvalho, and J. Ø. Nielsen, "Towards very large aperture massive MIMO: A measurement based study," in *Proc. IEEE Global Telecommun. Conf. (GLOBECOM)*, Austin, TX, Dec. 2014.

[47] X. Gao, O. Edfors, F. Tufvesson, and E. G. Larsson, "Massive MIMO in real propagation environments: do all antennas contribute equally?" *IEEE Trans. Commun.*, vol. 63, no. 11, pp. 3917–3928, Nov. 2015.

[48] H. Q. Ngo, E. G. Larsson, and T. L. Marzetta, "Energy and spectral efficiency of very large multiuser MIMO systems," *IEEE Trans. Commun.*, vol. 61, no. 4, pp. 1436–1449, Apr. 2013.

[49] H. Yang and T. L. Marzetta, "Performance of conjugate and zero-forcing beamforming in large-scale antenna systems," *IEEE J. Sel. Areas Commun.*, vol. 31, no. 2, pp. 172–179, Feb. 2013.

[50] S. Wagner, R. Couillet, D. T. M. Slock, and M. Debbah, "Large system analysis of linear precoding in MISO broadcast channels with limited feedback," *IEEE Trans. Inf. Theory*, vol. 58, no. 7, pp. 4509–4537, July 2012.

[51] R. Couillet and M. Debbah, "Signal processing in large systems: a new paradigm," *IEEE Signal Process. Mag.*, vol. 30, no. 1, pp. 24–39, Jan. 2013.

[52] J. Hoydis, S. ten Brink, and M. Debbah, "Massive MIMO in the UL/DL of cellular networks: How many antennas do we need?" *IEEE J. Sel. Areas Commun.*, vol. 31, no. 2, pp. 160–171, Feb. 2013.

[53] J. Jose, A. Ashikhmin, T. L. Marzetta, and S. Vishwanath, "Pilot contamination and precoding in multi-cell TDD systems," *IEEE Trans. Wireless Commun.*, vol. 10, no. 8, pp. 2640–2651, Aug. 2011.

[54] H. Q. Ngo, E. G. Larsson, and T. L. Marzetta, "The multicell multiuser MIMO uplink with very large antenna arrays and a finite-dimensional channel," *IEEE Trans. Commun.*, vol. 61, no. 6, pp. 2350–2361, June 2013.

[55] E. Björnson and E. G. Larsson, "Three practical aspects of massive MIMO: Intermittent user activity, pilot synchronism, and asymmetric deployment," in *Proc. IEEE Global Telecommun. Conf. (GLOBECOM)*, San Diego, CA, Dec. 2015.

[56] A. Lozano, R. W. Heath, Jr., and J. G. Andrews, "Fundamental limits of cooperation," *IEEE Trans. Inf. Theory*, vol. 59, no. 9, pp. 5213–5226, Sept. 2013.

[57] F. Fernandes, A. Ashikhmin, and T. L. Marzetta, "Inter-cell interference in noncooperative TDD large scale antenna systems," *IEEE J. Sel. Areas Commun.*, vol. 31, no. 2, pp. 192–201, Feb. 2013.

[58] I. Atzeni, J. Arnau, and M. Debbah, "Fractional pilot reuse in massive MIMO systems," in *Proc. IEEE Int. Conf. on Commun. (ICC)*, London, UK, June 2015.

[59] X. Zhu, Z. Wang, L. Dai, and C. Qian, "Smart pilot assignment for massive MIMO," *IEEE Commun. Lett.*, vol. 19, no. 9, pp. 1644–1647, Sept. 2015.

[60] J. H. Sørensen and E. de Carvalho, "Pilot decontamination through pilot sequence hopping in massive MIMO systems," in *Proc. IEEE Global Telecommun. Conf. (GLOBECOM)*, Austin, TX, Dec. 2014.

[61] E. Björnson, E. G. Larsson, and M. Debbah, "Massive MIMO for maximal spectral efficiency: How many users and pilots should be allocated?" *IEEE Trans. Wireless Commun.*, vol. 15, no. 2, pp. 1293–1308, Feb. 2016.

[62] W. Mahyiddin, P. A. Martin, and P. J. Smith, "Performance of synchronized and unsynchronized pilots in finite massive MIMO systems," *IEEE Trans. Wireless Commun.*, vol. 14, no. 12, pp. 1536–1276, Dec. 2015.

[63] J.-C. Shen, J. Zhang, and K. B. Letaief, "Downlink user capacity of massive MIMO under pilot contamination," *IEEE Trans. Wireless Commun.*, vol. 11, no. 6, pp. 3183–3193, June 2015.

[64] S. Jin, X. Wang, Z. Li, K.-K. Wong, Y. Huang, and X. Tang, "On massive MIMO zero-forcing transceiver using time-shifted pilots," *IEEE Trans. Veh. Technol.*, vol. 65, no. 1, pp. 59–74, Jan. 2016.

[65] H. Yin, D. Gesbert, M. Filippou, and Y. Liu, "A coordinated approach to channel estimation in large-scale multiple-antenna systems," *IEEE J. Sel. Areas Commun.*, vol. 31, no. 2, pp. 264–273, Feb. 2013.

[66] H. Huh, G. Caire, H. C. Papadopoulos, and S. A. Ramprashad, "Achieving "Massive MIMO" spectral efficiency with a not-so-large number of antennas," *IEEE Trans. Wireless Commun.*, vol. 11, no. 9, pp. 3226–3239, Sept. 2012.

[67] L. You, X. Gao, X.-G. Xia, N. Ma, and Y. Peng, "Pilot reuse for massive MIMO transmission over spatially correlated Rayleigh fading channels," *IEEE Trans. Wireless Commun.*, vol. 11, no. 6, pp. 3352–3366, June 2015.

[68] H. Q. Ngo and E. G. Larsson, "EVD-based channel estimations for multicell multiuser MIMO with very large antenna arrays," in *Proc. IEEE Int. Conf. on Acoustics, Speech, and Signal Process. (ICASSP)*, Kyoto, Japan, Mar. 2012.

[69] D. Neumann, M. Joham, and W. Utschick, "Channel estimation in massive MIMO systems." [Online]. Available: http://arxiv.org/abs/1503.08691

[70] J. Ma and L. Ping, "Data-aided channel estimation in large antenna systems," *IEEE Trans. Signal Process.*, vol. 62, no. 12, pp. 3111–3124, June 2014.

[71] R. R. Müller, L. Cottatellucci, and M. Vehkaperä, "Blind pilot decontamination," *IEEE J. Sel. Topics Signal Process.*, vol. 8, no. 5, pp. 773–786, Oct. 2014.

[72] D. Hu, L. He, and X. Wang, "Semi-blind pilot decontamination for massive MIMO systems," *IEEE Trans. Wireless Commun.*, vol. 15, no. 1, pp. 525–536, Jan. 2016.

[73] A. Ashikhmin, T. Marzetta, and L. Li, "Interference reduction in multi-cell massive MIMO systems I: Large-scale fading precoding and decoding," *IEEE Trans. Inf. Theory*, to appear. [Online]. Available: arxiv.org/abs/1411.4182

[74] L. Li, A. Ashikhmin, and T. Marzetta, "Interference reduction in multi-cell massive MIMO systems II: Downlink analysis for a finite number of antennas," *IEEE Trans. Inf. Theory*, to appear. [Online]. Available: arxiv.org/abs/1411.4183

[75] S. Lakshminaryana, M. Debbah, and M. Assaad, "Coordinated multi-cell beamforming for massive MIMO: A random matrix theory approach," *IEEE Trans. Inf. Theory*, vol. 61, no. 6, pp. 3387–3412, June 2015.

[76] M. Mazrouei-Sebdani and W. A. Krzymien, "Massive MIMO with clustered pilot contamination precoding," in *Proc. IEEE Asilomar Conf. Signals, Systems, and Computers*, Pacific Grove, CA, Nov. 2013.

[77] Z. Jiang, A. F. Molisch, G. Caire, and Z. Niu, "Achievable rates of FDD massive MIMO systems with spatial channel correlation," *IEEE Trans. Wireless Commun.*, vol. 14, no. 5, pp. 2868–2882, May 2015.

[78] J. Choi, Z. Chance, D. J. Love, and U. Madhow, "Noncoherent trellis coded quantization: A practical limited feedback technique for massive MIMO systems," *IEEE Trans. Commun.*, vol. 61, no. 12, pp. 5016–5029, Dec. 2013.

[79] J. Choi, D. J. Love, and T. Kim, "Trellis-extended codebooks and successive phase adjustment: A path from LTE-advanced to FDD massive MIMO systems," *IEEE Trans. Wireless Commun.*, vol. 14, no. 4, pp. 2007–2016, Apr. 2015.

[80] J. Choi, D. J. Love, and P. Bidigare, "Downlink training techniques for FDD massive MIMO systems: Open-loop and closed-loop training with memory," *IEEE J. Sel. Topics Signal Process.*, vol. 8, no. 5, pp. 802–814, Oct. 2014.

[81] X. Rao, V. K. Lau, and X. Kong, "CSIT estimation and feedback for FDD multiuser massive MIMO systems," in *Proc. IEEE Int. Conf. on Acoustics, Speech, and Signal Process. (ICASSP)*, Florence, Italy, May 2014.

[82] A. Adhikary, J. Nam, J.-Y. Ahn, and G. Caire, "Joint spatial division and multiplexing – the large-scale array regime," *IEEE Trans. Inf. Theory*, vol. 59, no. 10, pp. 6441–6463, Oct. 2013.

[83] J. Chen and V. K. Lau, "Two-tier precoding for FDD multi-cell massive MIMO time-varying interference networks," *IEEE J. Sel. Areas Commun.*, vol. 32, no. 6, pp. 1230–1238, June 2014.

[84] D. Kim, G. Lee, and Y. Sung, "Two-stage beamformer design for massive MIMO downlink by trace quotient formulation," *IEEE Trans. Commun.*, vol. 63, no. 6, pp. 2200–2211, June 2015.

[85] J. Park and B. Clerckx, "Multi-user linear precoding for multi-polarized massive MIMO system under imperfect CSIT," *IEEE Trans. Wireless Commun.*, vol. 14, no. 5, pp. 2532–2547, May 2015.

[86] C. Sun, X. Gao, S. Jin, M. Matthaiou, Z. Ding, and C. Xiao, "Beam division multiple access transmission for massive MIMO communications," *IEEE Trans. Commun.*, vol. 63, no. 6, pp. 2170–2184, June 2015.

[87] M. Matthaiou, C. Zhong, M. R. McKay, and T. Ratnarajah, "Sum rate analysis of ZF receivers in distributed MIMO systems," *IEEE J. Sel. Areas Commun.*, vol. 31, no. 2, pp. 180–191, Feb. 2013.

[88] A. Yang, Y. Jing, C. Xing, Z. Fei, and J. Kuang, "Performance analysis and location optimization for massive MIMO systems with circularly distributed antennas," *IEEE Trans. Wireless Commun.*, vol. 14, no. 10, pp. 5659–5671, Oct. 2015.

[89] M. Sadeghi, C. Yuen, and Y. H. Chew, "Sum rate maximization for uplink distributed massive MIMO systems with limited backhaul capacity," in *Proc. IEEE Global Telecommun. Conf. (GLOBECOM)*, Austin, TX, Dec. 2014.

[90] Y. Huang, C. W. Tan, and B. D. Rao, "Efficient SINR fairness algorithm for large distributed multiple-antenna networks," in *Proc. IEEE Int. Conf. on Acoustics, Speech, and Signal Process. (ICASSP)*, Vancouver, Canada, May 2013.

[91] J. Joung, Y. K. Chia, and S. Sun, "Energy-efficient, large-scale distributed-antenna system (L-DAS) for multiple users," *IEEE J. Sel. Topics Signal Process.*, vol. 8, no. 5, pp. 930–941, Oct. 2014.

[92] K. Hosseini, W. Yu, and R. S. Adve, "Large-scale MIMO versus network MIMO for multicell interference mitigation," *IEEE J. Sel. Topics Signal Process.*, vol. 8, no. 5, pp. 930–941, Oct. 2014.

[93] K. T. Truong and R. W. Heath, Jr., "The viability of distributed antennas for massive MIMO systems," in *Proc. IEEE Asilomar Conf. Signals, Systems, and Computers*, Pacific Grove, CA, USA, Nov. 2013.

[94] H. Q. Ngo, A. Ashikhmin, H. Yang, E. G. Larsson, and T. L. Marzetta, "Cell-Free Massive MIMO: Uniformly great service for everyone," in *Proc. IEEE Workshop on Signal Processing Adv. in Wireless Commun. (SPAWC)*, Stockholm, Sweden, June 2015.

[95] Y.-G. Lim, C.-B. Chae, and G. Caire, "Performance analysis of massive MIMO for cell-boundary users," *IEEE Trans. Wireless Commun.*, vol. 14, no. 12, pp. 6827–6842, Dec. 2015.

[96] J. Zuo, J. Zhang, C. Yuen, W. Jiang, and W. Luo, "Multi-cell multi-user massive MIMO transmission with downlink training and pilot contamination precoding," *IEEE Trans. Veh. Technol.*, to appear.

[97] A. Khansefid and H. Minn, "Achievable downlink rates of MRC and ZF precoders in massive MIMO with uplink and downlink pilot contamination," *IEEE Trans. Commun.*, vol. 63, no. 12, pp. 4849–4864, Dec. 2015.

[98] H. Q. Ngo and E. G. Larsson, "Blind estimation of effective downlink channel gains in massive MIMO," in *Proc. IEEE Int. Conf. on Acoustics, Speech, and Signal Process. (ICASSP)*, Brisbane, Australia, Apr. 2015.

[99] A. Kammoun, A. Müller, E. Björnson, and M. Debbah, "Linear precoding based on polynomial expansion: Large-scale multi-cell MIMO systems," *IEEE J. Sel. Topics Signal Process.*, vol. 8, no. 5, pp. 861–875, Oct. 2014.

[100] M. Wu, B. Yin, A. Vosoughi, C. Studer, J. R. Cavallaro, and C. Dick, "Approximate matrix inversion for high-throughput data detection in the large-scale MIMO uplink," in *IEEE International Symposium on Circuits and Systems (ISCAS)*, 2013, pp. 2155–2158.

[101] J. Hoydis, M. Debbah, and M. Kobayashi, "Asymptotic moments for interference mitigation in correlated fading channels," in *Proc. IEEE Int. Symp. on Inf. Theory (ISIT)*, Saint Petersburg, Russia, July 2011.

[102] H. Prabhu, J. Rodrigues, O. Edfors, and F. Rusek, “Approximative matrix inverse computations for very-large MIMO and applications to linear pre-coding systems,” in *Proc. IEEE Wireless Commun. Network. Conf. (WCNC)*, Shanghai, China, Apr. 2013.

[103] B. Yin, M. Wu, J. R. Cavallaro, and C. Studer, “Conjugate gradient-based soft-output detection and precoding in massive MIMO systems,” in *Proc. IEEE Global Telecommun. Conf. (GLOBECOM)*, Austin, TX, Dec. 2014.

[104] A. Liu and V. K. N. Lau, “Phase only RF precoding for massive MIMO systems with limited RF chains,” *IEEE Trans. Signal Process.*, vol. 62, no. 17, pp. 4505–4515, Sept. 2014.

[105] L. Liang, W. Xu, and X. Dong, “Low-complexity hybrid precoding in massive multiuser MIMO systems,” *IEEE Wireless Commun. Lett.*, vol. 3, no. 6, pp. 653–656, Dec. 2014.

[106] A. Alkhateeb, O. E. Ayach, G. Leus, and R. W. Heath, Jr., “Channel estimation and hybrid precoding for millimeter wave cellular systems,” *IEEE J. Sel. Topics Signal Process.*, vol. 8, no. 5, pp. 831–846, Oct. 2014.

[107] J. Zhang, X. Yuan, and L. Ping, “Hermitian precoding for distributed MIMO systems with individual channel state information,” *IEEE J. Sel. Areas Commun.*, vol. 31, no. 2, pp. 241–250, Feb. 2013.

[108] X. Li, E. Björnson, E. G. Larsson, S. Zhou, and J. Wang, “A multi-cell MMSE detector for massive MIMO systems and new large system analysis,” in *Proc. IEEE Global Telecommun. Conf. (GLOBECOM)*, Dec. 2015.

[109] ——, “A multi-cell MMSE precoder for massive MIMO systems and new large system analysis,” in *Proc. IEEE Global Telecommun. Conf. (GLOBECOM)*, Dec. 2015.

[110] S. K. Mohammed and E. G. Larsson, “Per-antenna constant envelope precoding for large multi-user MIMO systems,” *IEEE Trans. Commun.*, vol. 61, no. 3, pp. 1059–1071, Mar. 2013.

[111] ——, “Constant-envelope multi-user precoding for frequency-selective massive MIMO systems,” *IEEE Wireless Commun. Lett.*, vol. 2, no. 5, pp. 547–550, Oct. 2014.

[112] C. Studer and E. G. Larsson, “PAR-aware large-scale multi-user MIMO-OFDM downlink,” *IEEE J. Sel. Areas Commun.*, vol. 31, no. 2, pp. 303–313, Feb. 2013.

[113] J. Pan and W.-K. Ma, “Constant envelope precoding for single-user large-scale MISO channels: Efficient precoding and optimal designs,” *IEEE J. Sel. Topics Signal Process.*, vol. 8, no. 5, pp. 982–995, Oct. 2014.

[114] E. G. Larsson and H. V. Poor, “Joint beamforming and broadcasting in massive MIMO,” *IEEE Trans. Wireless Commun.*, vol. 15, no. 4, pp. 3058–3070, Apr. 2016.

[115] J. Zhu, R. Schober, and V. Bhargava, “Secure transmission in multicell massive MIMO systems,” *IEEE Trans. Wireless Commun.*, vol. 13, no. 9, pp. 4766–4781, Sept. 2014.

[116] E. Björnson, J. Hoydis, M. Kountouris, and M. Debbah, “Massive MIMO systems with non-ideal hardware: Energy efficiency, estimation, and capacity limits,” *IEEE Trans. Inf. Theory*, vol. 60, no. 11, pp. 7112–7139, Nov. 2014.

[117] E. Björnson, M. Matthaiou, and M. Debbah, “Massive MIMO with non-ideal arbitrary arrays: Hardware scaling laws and circuit-aware design,” *IEEE Trans. Wireless Commun.*, vol. 14, no. 8, pp. 4353–4368, Aug. 2015.

[118] U. Gustavsson, C. Sanchéz-Perez, T. Eriksson, F. Athley, G. Durisi, P. Landin, K. Hausmair, C. Fager, and L. Svensson, “On the impact of hardware impairments on massive MIMO,” in *Proc. IEEE Global Telecommun. Conf. (GLOBECOM)*, Austin, TX, Dec. 2014.

[119] A. Hakkarainen, J. Werner, K. R. Dandekar, and M. Valkama, "Widely-linear beamforming and RF impairment suppression in massive antenna arrays," *J. Commun. Netw.*, vol. 15, no. 4, pp. 383–397, Aug. 2013.

[120] S. Zarei, W. Gerstacker, J. Aulin, and R. Schober, "I/Q imbalance aware widely-linear channel estimation and detection for uplink massive MIMO systems," in *Proc. Int. Symp. of Wireless Communication Systems (ISWCS)*, Brussels, Belgium, Aug. 2015.

[121] L. Fan, S. Jin, C.-K. Wen, and H. Zhang, "Uplink achievable rate for massive MIMO with low-resolution ADC," *IEEE Commun. Lett.*, vol. 19, no. 12, pp. 2186–2189, Dec. 2015.

[122] C. Mollén, U. Gustavsson, T. Eriksson, and E. G. Larsson, "Out-of-band radiation measure for MIMO arrays with beamformed transmission," in *Proc. IEEE Int. Conf. on Commun. (ICC)*, Kuala Lumpur, Malaysia, May 2016.

[123] A. Pitarokoilis, S. K. Mohammed, and E. G. Larsson, "Uplink performance of time-reversal MRC in massive MIMO systems subject to phase noise," *IEEE Trans. Wireless Commun.*, vol. 14, no. 2, pp. 711–723, Feb. 2015.

[124] R. Krishnan, M. R. Khanzadi, N. Krishnan, Y. Wu, A. G. Amat, T. Eriksson, and R. Schober, "Linear massive MIMO precoders in the presence of phase noise – a large-scale analysis," *IEEE Trans. Veh. Technol.*, vol. 65, no. 5, pp. 3057–3071, May 2016.

[125] M. R. Khanzadi, G. Durisi, and T. Eriksson, "Capacity of SIMO and MISO phase-noise channels with common/separate oscillators," *IEEE Trans. Commun.*, vol. 63, no. 9, pp. 3218–3231, Sept. 2015.

[126] F. Kaltenberger, J. Haiyong, M. Guillaud, and R. Knopp, "Relative channel reciprocity calibration in MIMO/TDD systems," in *Proc. Future Network and Mobile Summit*, Florence, Italy, June 2010.

[127] P. Zetterberg, "Experimental investigation of TDD reciprocity-based zero-forcing transmit precoding," *EURASIP J. on Advances in Signal Processing*, Jan. 2011.

[128] R. Rogalin, O. Y. Bursalioglu, H. Papadopoulos, G. Caire, A. Molisch, A. Michaloliakos, V. Balan, and K. Psounis, "Scalable synchronization and reciprocity calibration for distributed multiuser MIMO," *IEEE Trans. Wireless Commun.*, vol. 13, no. 4, pp. 1815–1831, Apr. 2014.

[129] J. Vieira, F. Rusek, and F. Tufvesson, "Reciprocity calibration methods for massive MIMO based on antenna coupling," in *Proc. IEEE Global Telecommun. Conf. (GLOBECOM)*, Austin, TX, Dec. 2014.

[130] J. Choi, "Downlink multiuser beamforming with compensation of channel reciprocity from RF impairments," *IEEE Trans. Commun.*, vol. 63, no. 6, pp. 2158–2169, June 2015.

[131] H. Wei, D. Wang, H. Zhu, and J. Wang, "Mutual coupling calibration for multiuser massive MIMO systems," *IEEE Trans. Wireless Commun.*, vol. 15, no. 1, pp. 606–619, Jan. 2016.

[132] C. Shepard, H. Yu, N. Anand, L. E. Li, T. L. Marzetta, R. Yang, and L. Zhong, "Argos: Practical many-antenna base stations," in *Proc. ACM Int. Conf. Mobile Computing and Networking (MobiCom)*, Istanbul, Turkey, Aug. 2012.

[133] W. Zhang, H. Ren, C. Pan, M. Chen, R. de Lamare, B. Du, and J. Dai, "Large-scale antenna systems with UL/DL hardware mismatch: Achievable rates analysis and calibration," *IEEE Trans. Commun.*, vol. 63, no. 4, pp. 1216–1229, Apr. 2015.

[134] D. Liu, W. Ma, S. Shao, and Y. Shen, "Performance analysis of TDD reciprocity calibration for massive MU-MIMO systems with ZF beamforming," *IEEE Commun. Lett.*, vol. 20, no. 1, pp. 113–116, Jan. 2016.

[135] E. Björnson, E. de Carvalho, E. G. Larsson, and P. Popovski, "Random access protocol for massive MIMO: Strongest-user collision resolution (SUCR)," in *Proc. IEEE Int. Conf. on Commun. (ICC)*, Kuala Lumpur, Malaysia, May 2016.

[136] J. H. Sørensen, E. de Carvalho, and P. Popovski, "Massive MIMO for crowd scenarios: A solution based on random access," in *Proc. IEEE Global Telecommun. Conf. (GLOBECOM)*, Austin, TX, Dec. 2014.

[137] E. de Carvalho, E. Björnson, E. G. Larsson, and P. Popovski, "Random access for massive MIMO systems with intra-cell pilot contamination," in *Proc. IEEE Int. Conf. on Acoustics, Speech, and Signal Process. (ICASSP)*, Shanghai, China, Mar. 2016.

[138] X. Meng, X.-G. Xia, and X. Gao, "Constant-envelope omni-directional transmission with diversity in massive MIMO systems," in *Proc. IEEE Global Telecommun. Conf. (GLOBECOM)*, Austin, TX, Dec. 2014.

[139] M. Karlsson, E. Björnson, and E. G. Larsson, "Broadcasting in massive MIMO using OSTBC with reduced dimension," in *Proc. Int. Symp. of Wireless Communication Systems (ISWCS)*, Brussels, Belgium, Aug. 2015.

[140] H. V. Cheng, E. Björnson, and E. G. Larsson, "Uplink pilot and data power control for single cell massive MIMO systems with MRC," in *Proc. Int. Symp. of Wireless Communication Systems (ISWCS)*, Brussels, Belgium, Aug. 2015.

[141] D. W. K. Ng, E. S. Lo, and R. Schober, "Energy-efficient resource allocation in OFDMA systems with large numbers of base station antennas," *IEEE Trans. Wireless Commun.*, vol. 11, no. 9, pp. 3292–3304, Sept. 2012.

[142] E. Björnson, L. Sanguinetti, J. Hoydis, and M. Debbah, "Optimal design of energy-efficient multi-user MIMO systems: Is massive MIMO the answer?" *IEEE Trans. Wireless Commun.*, vol. 14, no. 6, pp. 3059–3075, June 2015.

[143] Y. Hu, B. Ji, Y. Huang, F. Yu, and L. Yang, "Energy-efficiency resource allocation of very large multi-user MIMO systems," *Wireless Netw.*, vol. 20, no. 6, pp. 1421–1430, Aug. 2014.

[144] H. Yang and T. L. Marzetta, "Total energy efficiency of cellular large scale antenna system multiple access mobile networks," in *Proc. IEEE Online Conference on Green Commun.*, Oct. 2013.

[145] S. K. Mohammed, "Impact of transceiver power consumption on the energy efficiency of zero-forcing detector in massive MIMO systems," *IEEE Trans. Commun.*, vol. 62, no. 11, pp. 3874–3890, Nov. 2014.

[146] T. M. Cover and J. A. Thomas, *Elements of Information Theory*. Wiley-Interscience, 2006.

[147] R. Gallager, *Information Theory and Reliable Communication*. John Wiley & Sons, 1968.

[148] B. Hassibi and B. M. Hochwald, "How much training is needed in multiple-antenna wireless links?" *IEEE Trans. Inf. Theory*, vol. 49, no. 4, pp. 951–963, Apr. 2003.

[149] M. Médard, "The effect upon channel capacity in wireless communications of perfect and imperfect knowledge of the channel," *IEEE Trans. Inf. Theory*, vol. 46, no. 3, pp. 933–946, May 2000.

[150] M. K. Varanasi and T. Guess, "Optimum decision feedback multiuser equalization with successive decoding achieves the total capacity of the Gaussian multiple-access channel," in *Proc. IEEE Asilomar Conf. Signals, Systems, and Computers*, Pacific Grove, CA, Nov. 1997.

[151] W. Yu, "Uplink-downlink duality via minimax duality," *IEEE Trans. Inf. Theory*, vol. 53, no. 2, pp. 361–374, Feb. 2006.

[152] H. Weingarten, Y. Steinberg, and S. Shamai (Shitz), "The capacity region of the Gaussian multiple-input multiple-output broadcast channel," *IEEE Trans. Inf. Theory*, vol. 52, no. 9, pp. 3936–3964, Sept. 2006.

[153] M. Costa, "Writing on dirty paper," *IEEE Trans. Inf. Theory*, vol. 29, no. 3, pp. 439–441, May 1983.

INDEX

Printed in the United States
By Bookmasters